GRANDPA'S LIES

Extraordinary Tales
of an
Ordinary Man

By
William A Easterling

ISBN-13: 978-0-578-70978-9

Dedication

This series of extraordinary tales would not be possible, were it not for the extraordinary support of my family. My wife, who has been by my side since I was 16. My daughter Lori and her husband Stew. My son Josh and his wife Annie. My grandchildren Avalyn, Kody, Harper, Kerrington, and Sam. Also, my parents, John and Lois, and their parents. My family has had a profound impact on every aspect of my life. With every step I made, it was my family that kept me on the path I blazed and motivated me to stay the course by supporting me in whatever adventure I undertook along the way.

I've written these memoirs with hope that this knowledge aides my children and my grandchildren to better understand not only who I was (and still am), but why or how some of my actions may have influenced who they are today.

Introduction

Extraordinary: going beyond what is usual, regular, or customary.

One Hundred Men, will Test Today, but only Three, Win the Green Beret.

Extraordinary.

My memoirs. My tales. My life. These stories form a collection of events that have marked up my memory with an indelible highlighter and include many portions of my past that encompass my time spent so far in this life as a husband, a father, and a warrior.

Rather than a true autobiography that could be expected from a historical figure where every instance of a famous life might be found entertaining, I have attempted to skip over the mundane, and focus on the more thrilling, the more exhilarating, the more extraordinary tales that oddly, seemed quite ordinary at the time.

My children grew up thinking that it was normal for dads to jump out of airplanes at night in groups of several hundred at a time. My wife considered it normal for her husband to leave for weeks or even months at a time with little to no information concerning where or why. But nothing about what happened was usual, regular, or customary.

My stories focus on various areas or times, and include military events such as parachuting, such as special operations training, and such as combat.

My stories also include memories of my youth and boyhood escapades that helped shape the man I would become. There are also accounts that show some of the more unique aspects of the lifestyle I shared with my family during the periods I was not serving in the US military.

Tied loosely together by the similarities of the tales, or the actual timeline in which they occurred, these revelations span the first sixty years of my life, and several hundred years of my family history. Included is as much truth as possible, based on memory, documentation, research, and family lore. On occasion, I've likely mixed in a bit of what could be called war-story content due to a shifting or cloudy memory from long ago. The combination should ensure that this memoir will be an entertaining, as well as an informative read. I do hope you enjoy it. I enjoyed living it.

Table of Contents

My Memoirs, Walking the Walk

I was born an American. My family has been American since before the Revolutionary War when our nation was the New World and still a British Colony. One of my direct ancestors supported the American quest for independence from British rule, and my family line has been serving America as military fighting men in all of our major wars ever since. Service to our Nation, as is found in so many families with deep roots in the Appalachian Mountains, seems to be a part of our genetic design.

I have spent considerable time and effort investigating my family history, our Easterling family history. I am including some of this information as a part of my memoirs, and while I can't guarantee that everything documented is 100% factual, it is the truth as much as I can prove, remember, or believe it to be. The stories contained here create my version of the truth as best I can present it, in what I'm calling, Grandpa's Lies.

Share my thoughts and memories as I provide an educated guess into the ancient history of the Easterlings of Athens County, Ohio, and El Paso County, Colorado. Likewise, I will share memories of my life from the time I can remember, starting at about four or five, and continuing through about my 60th year.

All knowledge is based on memories. Without them, we would start each day as a newborn babe. I hope my memories shared here will improve your life in some way.

Our Ancient Easterling History or Something Close

Ancient history for the commoner is rare as far as specific family lines are concerned. However, sometimes you find enough information to connect the dots and develop what is a good possibility of being a fairly accurate glimpse into the past. I have spent years, decades actually, doing sporadic searches and chasing down leads here and there. I'm quite confident that the history I have pieced together for this family story is as close to the truth as one could get without access to a time machine.

I've used multiple platforms to collect evidence and then I made fair assessments of how they could all tie together. I may not be able to prove that this is a perfect truth of the family's history, but neither can it be proven to be a complete falsehood. It's all backed up by documented information I have found in various genealogical repositories, a few personal stories collected from private sources such as Bibles and diaries or otherwise documented activities, as well as genetic markers and historical trails deriving from my DNA analysis. When it all is tied together, the story it tells seems to be quite believable.

The earliest reference to the surname Easterling, in what could easily be a direct ancestor, appears in some German genealogical libraries dating back to the early 800s AD.

Eyvind Austmann Bjornsson, born about 830AD in Norway, may be the first Easterling ancestor noted by the name Easterling. The name Austmann in this instance included footnotes in the documentation indicating that this was also someone known as "The Easterner" as translated to English. In at least one document, another variable of this name is listed as Eyvind The Easterling Austmann Bjarnasson.

This man's father was Bjarni Hrolfsson and his mother was Hilf. Eyvind sailed with his father in what is described as a "Viking warship" until his father died. The record states that after his father died, Eyvind, who now commanded the ship, likely a Longship or Karve, went on to raid locations in the British Isles. Eyvind married a woman around 845AD that he met, or perhaps captured, in the British Isles. She was named Raforta MacCearbhall, who was apparently a daughter from the Nordic family Kjarvald, as her maiden name is listed also as Kjarvaldsdottir. They had three sons, Oleg, Helgi, and Snaebjorn Eyvindsson, and two daughters, Thorid and Bjorg Eyvindsdatter. Eyvind is shown as having died in Iceland in or about 900AD.

The trail of Eyvind and his specific linage of family members sort of stopped here, at least as far as my limited research was able to show. Given the DNA evidence I discovered after having my own DNA analyzed, I believe that some of the ancestral Easterling family members remained in or returned to the United Kingdom and Ireland territories, eventually migrating eastward into Germany. It's also possible that some of the same family members returned to Norway and migrated south through Demark and on into Germany.

The early information, coupled with the DNA analysis, brought credence to some additional early history that I had gathered several years earlier. According to the DNA analysis, my genetic markers are 100% from northern Europe, the United Kingdom & Ireland, and the Scandinavian areas, with the remaining areas all being north of Spain and west of Russia. Likewise, Linda's DNA analysis, my wife and mother of our children, is almost identical to mine, with the exception of having more English markers and less German markers in hers. Given this physical evidence of the DNA, the documented historical information becomes much more conceivable.

Beginning in the early years of the 900AD timeframe, the surname Easterling has been found in the historical Viking Sagas and the Orkneyinga Sagas covering the Viking Age in and around northern Scotland and the Orkney and Shetland Islands. Another reference to the Easterling surname was found in the ancient records near Somerset, England as a part of the St. Donats Manor history.

Various documentation also shows that the name Easterling itself was changed by different spellings and even translations as time and locations of the family members changed through the centuries. Easterling has been recorded as Esterling,

Estarling, Eastarling, and even Stradling. However, our version of Easterling is shown to have remained as it is currently spelled, even from the time our family was living in Germany prior to the early 1700s. That being stated, some of the oral history passed down suggests the name may have been spelled, at least by some, as Osterling during periods of time in Germany, given that ost in German, translates to east in English.

So, as far as I can tell, this is a pretty complete and thorough history of our family as far back as can be expected to be found. For the most part, the stories ahead will be stories through which I have lived and remember as my own memories, or as discovered in written documentation from diverse American sources, to include the Sons of the American Revolution. This has been created with love for my children and grandchildren, and also in hopes that perhaps in a thousand years, my descendants won't have to guess what life was like for me and mine, while I walked the earth.

An Old and Narrow Lineage: Our American Background

The Easterlings of the United States began our history in the "New World" shortly before the Revolutionary War between the Colonies and Great Britain. Having arrived in North Carolina after travelling from Germany in the early to mid-1700s, the Easterling line is documented in multiple sources to include family histories, legal documents, and physical evidence such as tombstones and governmental records. One such document is my own certificate of membership in the genealogical organization the National Society of the Sons of the American Revolution, or NSSAR, via being a direct descendant of Thomas Easterling. Thomas Easterling is on the official US records showing that he had been paid for services rendered in NC during the Revolutionary War by the government of the United States. Thomas was born in NC around 1760 and died in Virginia on July 1, 1815. His father, the first American Easterling, was John. The family lineage is as follows.

John Easterling lived in the 1700s and was born in central Germany. Early records show that he migrated from near a village named Kultfabrik, nearby the town of Niederdoria, which is in the county or Stadt of Thüringentherme. This area is in the middle of Germany between Hamburg and München to the north and south, and Frankfurt and Berlin to the west and east. John married a woman named Eliza around 1729. Her formal name may have been Elizabeth as the diminutive version, Eliza became a popular nickname around the 16th Century.

John and Eliza had at least one child. Thomas J Easterling was either born in Germany about 1755 or in NC in 1760, depending on the source for the information I was able to find.

Thomas is the Easterling ancestor that is on the rolls as the family member listed for supporting the American Revolution, providing the qualification for membership in the Sons and Daughters of the American Revolution organizations for direct descendants.

Thomas married Rebecca Vickers and they had 11 children, one of whom was Silas Easterling. Silas is the first Easterling in our direct line that I could find a lot of information on as I did my genealogy research.

Silas Easterling was born about 1812 in Russell County, VA, and died on May 21, 1879 in Jackson County, OH.

Silas was born and raised in Russell County, VA, and then he moved to KY in 1834. This move occurred between marriages, perhaps after his first wife died. He apparently married for the second time, to his former wife's cousin, both with the last name of Morris. His second marriage was conducted in Beaver Creek, Floyd County, KY, on September 3rd, 1838. It appears that the Morris family also came from VA, from the region now known as West Virginia.

According to the 1850 census, they were living in Carter County KY but by 1860 had moved to Greenup County KY. For the 1870 census, they were listed as living in Jackson County, Ohio.

Silas served in the Union Army during the Civil War. He enlisted on March 2, 1863 and was a resident of Ohio when he enlisted. He served in Company K, 3rd Regiment, West Virginia Cavalry until June 30, 1865. He had two sons, John M. and Thomas E. Easterling who also served in the Union Army during the Civil War.

Silas was in his 40s when the war started, so I can presume that he must have had strong convictions about the war since he would not have been drafted or conscripted to fight for either side due to his age. Perhaps he joined because his two sons also put on the uniform of their state.

Silas' son, John M. Easterling, joined the Union Army on August 7, 1862 and was assigned to the 91st Infantry Regiment, Ohio Volunteers at the age of 24. He died from an accidental gunshot wound on May 27, 1865 in Depot Hospital, Winchester VA. His brother Thomas E. Easterling, who was also in the Union Army, was with him when he died.

Thomas was born in KY in 1843 and served with his brother John, after having enlisted on the same day and being assigned to the same unit. Thomas was discharged from the Army on June 4, 1865 only a few days after his brother John had died from his wound.

Another of Silas' sons, Allen O. Easterling is the next in our direct line. He was the brother to John M. and Thomas E. who fought in the Civil War. He was born in KY on October 4, 1848 and died in Carbondale, Athens County Ohio on September 30, 1922. He was married to Rachel Penrod who died in 1937. They had 12 children, one being John M. Easterling, which was the next one in our line.

John M. Easterling was born on July 5, 1875 and died in 1944. He married Sarah Melissa McClain on December 17, 1896. John and Sarah had 10 children according various official documents, as well as my father's personal memories. John and Sarah's children were named Allen, Chip, Jess, Glenn, Velma, Euma, Esther, Lloyd, Ruth, and Wylie. Allen Ona Easterling was the oldest of their children, and was also the next in our direct line.

Allen Ona Easterling was born on November 15, 1896 in Carbondale, Ohio, died on June 6, 1968 in Athens, Ohio, and was buried in New Marshfield, Ohio. He was married to Evelyn Price (maiden name) of New Marshfield, Ohio. They had two children, Esther and John Wesley Easterling, my father.

John Wesley Easterling was born in Athens, Ohio in 1935 and married three times. He and his first wife, Lois Marie Perry (maiden name) were my parents. He was later married to Sally, and then Patricia.

Of course, next in line is me, myself, and I. I was born William Allen Easterling in Athens, Ohio in 1956, at the Sheltering Arms Hospital. I married Linda Lue Elkins (maiden name) in 1975. We had two children: Lori Lynn, born in 1978 at Ft. Carson Army Hospital, Ft. Carson, CO; and Joshua Joseph, born in 1984 at the US Army Hospital, Augsburg, Bavaria, Germany.

Lori married Stewart McDowall in 2007 in Colorado Springs, CO. They have two children as of the writing of this story, Kody Aidan and Kerrington Isla.

Joshua married Annie Miller, also in 2007, in Colorado Springs, CO as well. They have three children as of the writing of this story, Avalyn Marie, Harper Irene, and Samuel Joseph William.

A Story of Silas Easterling, my Great-Great-Great Grandfather

Silas Easterling hailed from Virginia during times before the Civil War. From what can be found in historical records and some family history, like most of the Easterling men in my direct-line, he was a coal miner. It's a fair bet that the Easterling family that sailed across the ocean from Germany before the Revolutionary War took place, also mined coal in the former fatherland. The relevance of this information is that in starting out as a Virginian, our branch of the Easterling tree would have been growing from western Virginia, nearer the coal fields that extend across the coal country of Pennsylvania, the Ohio River Valley region of Ohio and (post-Civil War) West Virginia, and on into Kentucky.

Once the War Between the States started, the western part of Virginia broke away and became its own state of West Virginia. Because of this, there were actually military units from West Virginia fighting on both sides during the Civil War.

Silas was a member of the 3rd West Virginia Volunteer Cavalry Regiment. This was the first Union Cavalry unit formed in West Virginia and the first company was recruited and organized from Morgantown in December 1861, under the Union government HQs located in Wheeling.

According to Civil War pension records, Silas was a trooper, but perhaps not of the highest caliber. He was old for a new volunteer, in his early 40s, and had several periods of his time in the military listed as missing from the ranks. He apparently participated in a battle known as the "Skirmish of Hurricane Bridge" that was fought over several hours on March 28, 1863.

The Union forces were positioned near the bridge crossing the creek and were about 150 strong with members from the 13th Volunteer Virginia Infantry and the 3rd West Virginia Volunteer Cavalry. The Union force was met by a Confederate force of over 500 as they marched towards Point Pleasant to attack a Union fort. The Union forces fought hard and although outnumbered and surrounded on three sides, they held their ground and forced the Confederates to pull back and bypass their position as they continued towards Point Pleasant.

It was after this skirmish that Silas came up missing. He may have been sent after the Confederate soldiers once they departed the area as a member of the Cavalry. It was quite common for men to be listed as missing after a major battle. Sometimes they would get lost in the woods or end up with a different unit. Some would take the opportunity to desert their unit and go home, at least for a while. Silas showed up on the military roles again, about three months later at a remount camp. This was a place set up for Cavalry troopers to go and get a new horse and equipment once their own mount was wounded or killed.

Some have speculated that he spent time near Parkersburg and may have been staying in the area to visit friends or relatives as he took a break from the war. He finished up with his unit, and at the end of the Civil War, he was discharged and went back home. Regardless of what happened during the Skirmish at Hurricane Bridge and for a few months after, he completed his tour of duty in various other places to include the Battle of Gettysburg and General Lee's surrender at the Appomattox Court House. The regiment also participated in the Grand Review of the Armies in Washington, DC and was mustered out on June 23, 1865.

Branches Grown from Deep Roots
My Paternal Line

Every family has secrets. Mine is no exception. My dad and I look like twins, separated by 15 or 20 years, so the bloodline is evident. Beyond that, well, some tales will be told and some will remain in the annuals of yesterday, soon to be forgotten.

For instance, my mom's second husband is an actual family-member, by blood. Take a journey with me as I travel back to the 19th and first half of the 20th Centuries and peek at life for my family ancestors who lived in Athens County, Ohio.

When Ohio was being settled, one of the founding families was the Haning family. Dale Haning married my mom several years after she and my dad had gotten divorced upon his return from Vietnam in 1969.

Back in the 1800s, Athens County, Ohio was home to not only the Easterling family, but to the Price and Haning families. The Haning family is one of the Athens County Historical Society's official First Families of Athens County, Ohio.

Dale Haning, my mom's 2nd husband and therefore my stepfather used to say that if your family was in Athens County before the 20th Century, you were probably a relative of his. The Haning family was once quite well off and had donated land to the Ohio University and even had a large university building in Athens named Haning Hall, after the Haning family. My stepfather, however, was not exceptionally wealthy, nor did his family still have those kinds of riches during his, nor the preceding generation of Hanings who were still living in Athens County.

In the early 1900s, my grandfather Allen O Easterling married my grandmother Evelyn Price. Her father, James W Price, married Belinda E Haning, thus linking the Easterling and Haning families together. This made Belinda E Price (Haning) the great-granddaughter of Matthew Haning Sr. Matthew is one of the founding family members listed with the Athens County Historical Society. He was born in 1730 in Germany and migrated to Athens County, Ohio, where he died in 1807.

So, between me, my dad, and my stepdad, we were all also cousins of some sort, by marriage and by blood.

My Maternal Line

My mother was born to Iscle Perry (Fierce) and William Perry, a coal miner with the former Luhrig Coal Company.

My maternal lineage, like my paternal line, also has deep roots in Athens County, Ohio.

My mother's mom, Iscle Marie Perry (Fierce) was the daughter of a local farmer, James Harvey Fierce. James owned a large farm just outside of Athens, Ohio in an area formerly known as the Luhrig Community, and lived in one of the first large brick and stone homes built in Athens County. He was born on April 29, 1876 and died on May 28, 1962 in Athens County, Ohio. He was married to Mary Elizabeth Robinson.

Mary Elizabeth Fierce (Robinson) was born on May 12, 1877. She was the daughter of Francis C Robinson, born October 2, 1831, and Mary Jane Johnson, born April 22, 1834. Both were born in Ames Township, Athens County, Ohio. Mary married James and moved to Athens from the Glouster, Ohio area, about 20 miles north of Athens County. The Robinson family owned farm property in the Glouster area, just north of Ames Township, Athens County, Ohio.

James Fierce's father, Wayne Dodson Fierce, moved to Athens County from an area near Cleveland, Ohio called Fierce Ridge, located in Trumbull County, Ohio, about 20 miles from the shore of Lake Erie.

William Perry, my mother's father, was born and raised in Athens County. His father, Jasper N Perry, also called "Gyp" or "Hap", was said to be a preacher, possibly of the Church of Christ. He married Lena Barnes from New Marshfield (Athens County), Ohio. Jasper's father was George T Perry, and his father was William Perry, born in the state of Delaware in 1798, and then moving to Athens, where he died on April 21, 1868.

My grandmother, Iscle Marie Perry (Fierce), had a younger sister named Dorothy Elberta Fierce (Aunt Dot). She married a local man named Luther Eugene "Gene" Yoho. He founded the L.E. Yoho Construction Company. My dad was working for the

Yoho Construction Company when he joined the US Army as a full-time soldier, having spent some time already as a National Guardsman with a unit from Athens, Ohio.

The Easterling family lines from both the paternal and maternal side run deep in Athens County, Ohio and the city of Athens, with the roots including the Easterling, Perry, Price, Fierce, Haning, Robinson, and other prominent families of the region.

My Parents and a Story

The mid-1950s had a different aspect on what was, or was not normal or accepted, regarding social behavior and activities. Social clubs were a great part of life in that time. Most homes had no television and many people were limited on resources or funding to be used for recreational events. The popular alternatives for activities other than work-related functions were churches and clubs.

Social clubs, such as the Eagles, Moose, or Odd Fellows lodges, were venues that allowed people to gather together for entertainment and recreation. Dances, celebrations, and other happenings, along with drinking in a bar or dining type of atmosphere was a very acceptable notion for weekly gatherings. Entire families would attend many of the events.

My mom met my dad at such an event. On a dare, according to my dad, he got my mom to agree to dance with him at the urging from one of his friends. Also, according to my dad, my mom looked several years older than she actually was, which helped explain the difference in their ages. She was young by 21st Century standards, being about 13 years old. My dad, just over four years older, started dating her and both sides of the family were not only aware of the relationship, they encouraged and supported it.

Within a year, in May of 1953, they were married and had a baby on the way. The baby, a boy, was stillborn and a couple of years later, I was born. I like to believe, and I've told my dad this story, that I am the person of the soul that started out with the first baby. I just had to wait a little longer before I left Heaven to get my start on life as a part of this family.

My mom was still practically a child when I was born, having just turned 17.

It might have been their ages, it might have been something else, but after I was born, my mom and dad had a very bumpy marriage over the next 13 years. Still, in spite of the difficulties, they loved each other when they were together and starting out as kids, and growing together into adults, there were likely more good times than bad. However, after 16 years of marriage, it was over. The divorce that took place once my dad returned from Vietnam was best for all concerned and they remained mostly friendly until my mom died in 2009.

The Halves

Long story short, my mom and dad had one child. I am the one son. While they were still married, my mom had three more children, albeit, not by my father. The two girls were raised as my sisters and my dad is listed as their father on their birth certificates. The baby boy was put up for adoption when he was delivered.

As I've stated, while my parents were together, they did well enough and managed to enjoy life and make life enjoyable for those around them, including the kids. While they were apart, like summer camp for the National Guard, like the year my dad was in Korea, or the long periods of time he was downrange during our trip to Germany, or while he spent his tour in Vietnam, my mom missed having adult companionship and made up for it with paramours who didn't mind that she was married.

It didn't make her a bad mom per se, but it tended to rub against the norm for acceptable conduct in a traditional marriage. Having an insider's access to some of the family secrets, I know my dad was not always a saint while he was gone, but he did manage to keep most of his mistakes away from his own backyard. Some will proclaim that this is a sexist thing to assert, especially since his job took him away and a stay-at-home housewife's role prevented such opportunities. But, this changes nothing about what has happened in the past and it remains a part of my story.

My dad ended up married again after he and my mom got divorced. His second wife, Sally Easterling (Hammock), and he produced two more half-siblings to me. My dad and his second wife also eventually divorced, and he soon married his third wife, Patricia.

My mom also married again and she and Dale Edward Haning had a son who became another one of my half-brothers.

In total, I ended up with three half-sisters (two from my mom, one from my dad) and three half-brothers (again, two from my mom, one from my dad).

As I grew up, I lived a life much different than any of my half-siblings. Due to physically being in different areas of the country, as well as having very different lifestyles, after the age of 16, I rarely had contact with any of them.

Our lives, having been created close together physically, were pulled apart by the choices we each made, good and bad. We all became adults and chose which paths to travel as we set out on our own life's journeys. Thus showing, that a common starting point is no assurance of a common destination point.

And yet, I am at peace with all that has transpired and why. As we've all heard before, life is like a box of chocolates...you never know what you're gonna get.

When I Was One

One of my favorite poems that I learned late in life was written by someone named Milne. Titled, “Now We Are Six”, it was not about me, because I was not six, but was close to 60. This poem was one that my grandchildren, first Kody, and then Kerrington, memorized as a part of their home-schooling curriculum. I liked the poem, but love the memories it now provides as all my grandchildren have grown beyond six and race quickly towards being grown-up. But, I relish the words as they always provide me with a good set of memories.

When I was One,
I had just begun.
When I was Two,
I was nearly new.
When I was Three
I was hardly me.
When I was Four,
I was not much more.
When I was Five,
I was just alive.
But now I am Six,
I'm as clever as clever.
So I think I'll be six now, for ever and ever.

Where do children go when they are grown up? They are still there, inside the grown-up, but sometimes they are hard to see. Still yet, especially for parents, the child in the grown-up is even harder not to see.

Some of us grownups are lucky because we also get to see the children who have grown up into wonderful adults, return as grandchildren who rightly or wrongly, bring with them opportunities for certain do-overs, or being able to complete some events that were once missed as life in all its glory, created more to do at any given time than could be done.

I was once one, but I don’t remember it. I’ve been told many stories about my first few years though. I was apparently born with a twisted neck, which required massages quite often as a brand new baby. I’m told it was a painful experience for both me and my mom or dad as they held me and twisted or turned my neck to get my muscles to relax.

I also had a dog named Skippy, and as a toddler, perhaps at the one-year point, I decided to bite the dog on the tail. As someone older would expect, but I'm sure for me, it came as quite a shock, Skippy turned around and bit me back. Near my eye, no less. My Grandma Perry, my mom's mom, must have taken me to the doctor, at least that's the way it looked in the old pictures I've seen. But yes, in this collection of extraordinary tales, I am one of the people who bit a dog and was bitten in return.

Movie Memories

One of my first actual memories came around the time I was four, I believe anyway. We all lived in a house just behind the Skyline Drive-in Theater in Elizabethtown, Kentucky, or E-Town as we called it. My dad was stationed with the US Army at Ft Knox at the time.

I remember seeing the Mickey Mouse and Donald Duck cartoons on the movie screen from our house porch. At least I seem to remember seeing them that way. I know we also drove to the movie and I remember playing on the swings beneath the screen before the movie would start. I'm pretty sure I was wearing my PJs too, which I thought was neat. Swinging on the big swings beneath the screen as the movie projector shined advertisements for the snack bar food and drinks, was a lot of fun.

I remember returning to this drive-in after we had moved onto the Army Post of Ft Knox and watching a film called Fantastic Voyage. Even as a young boy, I enjoyed watching the starlet in a skin-tight SCUBA suit swimming around in someone's blood stream. The concept of shrinking a submarine full of doctors to go inside a body to treat an illness was ok, but the large screen vixen was even better. I continued to enjoy drive-in theaters for many years in several states. They may even make a big comeback one day.

Raccoon Camp

My next stage in life, when I can really start to recall memories was while we lived in Athens, Ohio, where both of my parents and I had been born. After living in E-Town and Ft Knox, my mom, sister Cathy, and I moved to Athens, while my dad was sent to Korea for an unaccompanied tour of duty.

I was five at the time and I have a memory of being asked if I would like to go to Kindergarten or stay home and wait until First Grade to start school. I opted to stay home and never went to Kindergarten. I also remember one of my favorite toys, a tan colored, metal pickup truck. I also have memories of an unsuccessful attempt to camp out in my backyard. I really don't know how much is the actual memory of the event or how much is a memory of being told about what happened, but I'm sure between the two, I have a good idea what transpired.

I had set up a little pup tent in the back yard. It might have been an Army pup tent that my dad had brought home, or it might have been one I made with a blanket and a couple of sticks. It was a tent, nonetheless. I had collected a jar full of lightening bugs and it was full. Holes having been poked in the lid, there might have been a hundred or more flashing bugs flying and crawling around inside what was probably a pint or quart sized Mason Jar.

I had hauled some blankets and pillows, along with a few stuffed animals out to the tent, making several trips in the process. My Grandma Perry and Grandpa Slats were there. My Grandma Easterling was probably there, as she spent a lot of time with us and usually if you saw one Grandma, you seen them both as they were not just relatives, they were best friends.

At some point, just before, or as it was starting to grow dark, Grandpa Slats started to tell me all about how a raccoon lived around there and would almost assuredly come to visit me during the night.

The idea of the raccoon coming into my tent in the night must have been more than I was willing to accept. It was about then that I grabbed everything I had spent several trips carrying outside to the tent, and in one trip, carried at all back into the house, saying as I went, "I ain't gonna sleep with no coon."

I'm pretty sure the next time I camped out was with my dad in what would have been several years later as we explored the wilderness of Ft Knox once he was assigned there again in the future.

Rammersdorf Rat, Munich, Germany, 1962 – 1964

Back in the early 1960s, a popular style of movie was the teenage beach party genre. Rock and Roll, surfing, and of course bikinis were all the rage. Even though I was just a kid, I still found some ideas in these movies I enjoyed, although I wouldn't admit that one might be the girls. One of these ideas I did like, was known as The Rats.

The Rats was the name of a goofy gang of motorcycle riding hoods who bothered the beach bums and bunnies as they had their adventures on the silver screen. Naturally, I thought it would be much cooler to be a Rat than to be a surfing teenager going gaga over the girls. I was only 7 or 8 and girls had not been elevated to my radar yet, except as creatures bearing cooties or something like that. Anyway, that's what I claimed.

The guys in the movies all had leather motorcycle jackets with a big rat on the back. My dad couldn't find me a black leather jacket, but he did take one of my coats and drew a replica of the movie version's gang logo of the Rats jacket on my coat. Having my Rat Jacket coat made me the natural leader of my own gang of Rats as we played in and around the military housing area.

We were small terrors as we rode roughshod over the streets of Rammersdorf on our bicycles. We would venture out of our neighborhood and carrying a section of a bamboo cane pole normally used for fishing, we would zip past the German kids and smack them on the legs. Worse yet, we'd ride down the side streets where we would find parked Mercedes sedans and snatch off the famous star hood ornaments as trophies of our conquests.

Many times, as I walked to and from the bus stop for school, I would pause long enough to have a fist fight with some kid who had dared to do something I didn't like. Fighting seemed to be a normal thing to do, and while not everyone had fights, I certainly had my share. I remember one fight in particular because I'd put a lot of thought into it before it ever happened.

One of the boys who lived in another apartment building had been picking on my sister and she had come to me complaining about it. His name was Ray and he was older and bigger than I was, but I couldn't let him get away with bothering my little sister. I remember talking to a couple of grown-ups about it while I was hanging around the apartment complex parking lot for families who lived in our building. They might have been teenagers or actual adults, but in my mind, they were grown-ups.

I remember one of them telling me how some kids are bullies and will pick on girls or younger kids, but that they were all really cowards. This sounded like some true words of wisdom and I continued to listen to his advice on the matter.

He also told me that I should just use surprise as a way to get the advantage against this kid who was older and bigger. So that's what I did.

I went over to Ray's street and found him playing outside. I starting talking to him and joking about how he had been bothering my sister. I acted like I thought the whole thing was funny and I was even smiling as he admitted to what all he had been doing. I was still smiling as I punched him right on the nose, knocking him down in surprise and pain. While he was down, I followed the sage advice I'd been given, and started kicking him a few times until he started to cry and ran for his house. I had acted in true Rat Pack fashion and beat up my foe.

What I hadn't considered was that for the next few weeks, I had to watch over my shoulder for any retribution until things cooled back down.

My Rats and I continued to run wild and do the things that young boys do when they are only minimally supervised. We played on a local construction site where new apartment buildings were being built, even as we would be chased off by both the workers and occasionally security guards.

There were still numerous buildings that had been bombed during WWII in our area and we would explore those, sometimes crawling into a bombed building's basement and finding some piece of Nazi memorabilia in the form of a coin or in one instance, an old gas mask. We were chased away from a bunker that we had been playing in once by some German with a long gun. He didn't shoot at us, but we were certain that he was going to, and also that he must have been an old Nazi.

Often, during the hours of darkness we would venture into a nearby park and capture hedgehogs. These were small animals that looked like little porcupines and we would keep them in boxes for a few days as pets until we'd release them back into the wild. The wild areas of the park anyway.

Comic books were king makers during the time I was in Germany as a kid. There was no American TV, so we read a lot of comic books. My favorites were (obviously) the military comics. SGT Rock, GI Combat and the Haunted Tank, SGT Fury and his Howling Commandos, and Our Army at War were my favorites. But, I also liked Archie, some of the Classics Illustrated, Beetle Baily, Sad Sack, Betty & Veronica, Richie Rich, and whatever scary comics I could find.

I would actually carry a bag of comics around the neighborhood, knocking on the apartment doors seeing who wanted to trade comics. It was a different and safer world back then. Sometimes, I'd find someone who would want to trade a more adult-version of a comic, like a Mad Magazine or one filled with adult cartoon jokes. These were always worth several normal comics, and hard to find.

Although, some of the better girly cartoons could be found in the Army's PS magazine, a Preventative Maintenance monthly magazine, featuring the blond bombshell Connie Rodd. I literally had hundreds of comic books when we moved back to the US. If I had kept them until I grew up, they would have been worth a small fortune.

The winter was a special time. There were two big hills that I would go ride my sled on when it snowed, and it seemed to snow a lot. There was a nice smooth hill near the German Park and a little farther away, there was another hill we called the Monkey Hill. I have no idea why, but that's what I remember it being called. It had a lot of large rocks on the top and had trees growing on the hill that gave us obstacles to avoid.

The standard American Flyer was the favored sled of the sledding community. We would rub candle wax on the runners and when the snow got really packed down, you would fly, sometimes literally, down the hillside and over the bumps. You could steer them with your hands while laying down, or your feet if setting on them. Unless you had a rope tied to the handles and then you could still use your hands.

The Germans had sleds made differently and they were taller, more like a horse drawn sled, and harder to steer. You had to lean to steer, or drag a foot on one side or the other. Like I said, the American Flyer was the best choice and a lot of fun was had on those two hills.

I have to mention one more memory of being in Munich as a kid. It's where I discovered the Super Ball. The super ball was a hard rubber ball about the size of a tennis ball. While common now, they were brand new, or new to me, back then. You could buy them at the German stores just outside our neighborhood for one or two Deutsch Marks, or back then, between 25 to 50 cents. You could throw one hard into the street and it would bounce to the top of the three-story apartment buildings. The super ball seemed like magic.

This was also the time that Barry Sadler's hit, Ballad of the Green Berets came out. It was played a lot on the Armed Forces Radio Network and I loved the story it told. I am told that my mom didn't like the song and didn't like the fact that I was even listening to it. My dad, on the other hand, liked the song as much as I did.

No one had any idea at the time that I would grow up and earn my own Green Beret, but it was a part of my earliest memories and always reminds me of living in Munich as a kid.

Driving Tanks and Playing Army, Henry Kaserne, Munich, Germany

Being a kid in Germany wasn't all fun and games. Being the son of a Tanker who was a Platoon Sergeant in an Armor Company came with a special set or responsibilities. One of those was getting to learn how to drive a behemoth. That behemoth being, an M60 Tank.

Costing about a half-million dollars during the early 1960s and weighing in at just over 50 tons of deadly combat power, the M60 "Patton" tank fit the bill for behemoth. My dad had his own tank and supervised the operations of another four tanks in his platoon. He and his tank crew of three other men had recently earned the title of Top Gun for the entire Battalion, and his platoon had earned the highest score of any other platoon during gunnery competitions.

I had my own set of Army fatigues which I wore whenever I went to work with my dad. I also wore the same rank on my fatigues and I was the smallest Staff Sergeant to get promoted to Sergeant First Class (Platoon Sergeant) on the entire Kaserne. Of course, my dad's office was in the Motor Pool, the parking lot for the tanks and other military vehicles. It was here, where I learned as much about the M60 tank as any eight-year-old should be expected to learn. I was the only kid around that knew not only how to start the tank, but to put it into gear and drive it around, which I did on occasion in my dad's Motor Pool.

During a special event, a family day or some type of celebration, the Battalion conducted a drill team competition and of course, my dad's drill team won. They marched and manipulated their M1 rifles while wearing their dress uniforms. It was quite impressive. There were also static displays of the vehicles, which obviously included the war horse, the M60 tank. My dad's tank, much to my frustration, was named Cathy's Clown, after my sister Cathy and a popular song on the radio at the time. I would have named it after myself of course. But, being in Charlie Company, they followed the naming convention for the tanks by having the names start with a C for

Charlie Company. The name was painted on the side of the 105mm main tank gun.

At the end of the celebration I was able to drive the tank back to the motor pool. I even recall having to drive around a parked vehicle that was in the street. To me, that was a huge demonstration of my tank driving skills. There was another time when I went to the field, a local training area, with the entire company of tanks. I rode on the outside of the turret of my dad's tank until it was time for the tank crew to run their drills. During the drills, I was able to watch safely from a distance, as the tank crew went through its paces, practicing the art of tank warfare.

For a young lad, these were awesome experiences and pretty much set the mold for my future longing to join the Army as a Tanker. That didn't happen though, and I ended up starting out in the Artillery instead, before volunteering a few years later for training to become Special Forces qualified. But, for a little while, I was a Tanker and I drove a tank.

The Red Clay Hill, Ft Knox, KY 1965 – 1968

Returning to Ft Knox after spending our tour in Germany was great. I wasn't going back to old friends, but I was going back to America and TV, and a place where I got to see my dad a lot more as well. At Ft Knox, my dad would come home from work almost every night instead of always being gone like he was in Germany. This allowed us to do a lot more together during what ended up being the last few years we would have due to changes that would soon impact our lives.

But during the weekends when I was not in school, and every day during the summer when I could go outside and play, the Red Clay Hill called to me.

Ft Knox is built on red clay. And while the clay may have been hard to wash off the various tanks and trucks at the wash racks around the motor pools on post, the clay was fun to play on. When you had enough rain, the clay hill could be transformed into a toboggan run and allow someone willing to get very dirty to slide down a cliff and into a big pool of ooze and goo at the bottom.

Often, my mom would have to turn the hose on me when I got home before I could even think about coming into the house and getting a bath. During dry weather, the clay would transform into a few inches of red dust and provide a perfect place to play construction, road building, or of course, Army. At the top of the hill was the woods. I built many forts and climbed many trees in those woods. At the bottom of the hill was a big ditch and there were a couple of large drainage pipes to catch the water flow from the hillside and direct it off towards some creek out past the highway across the field. Of course, they were perfect places to explore, at least in my mind. One of the large pipes passed beneath a roadway and when you were in it, you could hear the cars running over the top. No one in authority ever had any idea that the buried culvert pipes were also a part of my extended playground. But I always had enough sense to stay out of them if there was any sign of mud or water inside the pipes.

At one end of the field running along the entirety of the hillside, was a small store. Something like a convenience store, not a super market. If I had a little money, I'd go into the store and buy the cheese and cracker packs.

Out behind the store and nearer the wood line, was an area covered in huge boulders. They were not natural, but had probably been dumped there during construction in the area as they were moved out of places where buildings were built and roads or highways had been put in. Whatever they were, it was all a perfect place to play Army. You could climb mountains, explore caves, find secret tunnels, or jump from the tops of hideouts to capture the enemy. Looking back now, I am surprised I never ran across a snake or even got hurt playing. Perhaps the angels were watching over me back then. Neither the red clay hill, nor the piled-up boulders would have been on my list of acceptable playgrounds for my own kids or grandkids today. But that was then, this is now. Life was different back then.

Sometimes Soldiers Die

The sky was clear and the sun was bright. My dad had taken me downrange to watch a capabilities and firepower demonstration being put on by one of the Ft Knox military organizations. There were static displays where we could see some of the equipment up close and even climb around on some of the vehicles.

There were also wooden bleachers standing along the edge of a long, gently sloped range that led down to the impact area. At one point, the loudspeakers directed all the onlookers to move to the viewing stands and my dad and I took our places somewhere near the top few rows of the bleachers.

We could hear the helicopters before we could see them. Flying in from the impact area, two UH-1H "Huey" Iroquois helicopters were headed directly towards the bleachers. Each was loaded with soldiers and the plan was to fly near the stands and perform a tactical dismount and security perimeter, followed by a patrol formation and movement out of the area as soon as the choppers quickly flew away. This was being done to show the people in the stands an example of what was happening in Vietnam as the war there was in full swing. Today, however, the violence of war would be brought much closer to the people watching this presentation unfold.

The sortie, the designation for a flight of aircraft flying a particular mission, was flying one behind the other with the trailing helicopter slightly offset from the lead bird. They were flying low, maybe 100 feet or a little more off the ground. As they grew closer to the stands, still flying at normal speed, not yet starting to get ready to land, the two choppers flew too close to each other and the main rotor blades crashed together.

When the large main blades of the helicopters hit, the one behind the lead bird immediately dropped to the ground and exploded into a huge fireball.

The people in the stand gasped, yelled out, or set motionless in disbelief as the tragedy unfolded before everyone's eyes.

The lead chopper began to shake and swing back and forth as the pilots fought for control. It continued to fly towards the bleachers, swinging and swaying like a drunken dragon on attack. My dad grabbed me and was getting ready to drop me beneath the bleachers in an attempt to escape what seemed to be a quickly-arriving catastrophe. And then, the 2nd chopper crashed into the ground.

The second bird didn't immediately explode, and lay there smoking as some of the soldiers began to clamber out of the wreckage as it also began to catch on fire. I had just been quickly, and in a frightening way, introduced to just how fast things can go wrong and bring unexpected death and destruction into the middle of an otherwise normal day.

Several soldiers died in front of my eyes that day and while not in a large way, it did change me in that instant. I believe, witnessing the crash instilled an extra amount of caution into my personality that has always remained.

The brutal reality is that in the military, primarily in the direct combat arms positions, training not only produces stress, sweat, and tears, it often-times brings injury and death. For me, this horrible accident became a teaching point for me as a future soldier and father.

I wasn't traumatized by the event, but it did help me understand that death is real and sometimes soldiers died.

Wading Creeks and Camping on a Special Island

I was too short to wade through the deep holes unaided, but my Army back pack kept me afloat as I followed my dad up the creek and through the woods as we searched for the perfect campsite.

Camping along a creek running through down-range training areas of Ft Knox, Kentucky was a trek into a world of past and present, mixed with a dose of future. It was a place of as yet, undetermined importance. A portal that had opened and would one day close as it tied both ends of a long journey into a complete circle.

The creek was clear and the water cool. My dad would park the car off the side of the road and we would grab our gear from the trunk, hang the backpacks off our shoulders, and start into the darkened woods following the sparkling path filled with water running to meet us and then flowing away behind as it continued its journey to the Ohio River and eventually, the Gulf of Mexico.

During our walks along this creek sometimes we would find signs of the Army which had used this area as a practice battleground in countless practice wars. A short belt of machine gun ammo here. A tank's track pad there. A rusty C-Ration can near the base of an old tree. A tangle of concertina wire grown over by briars and blackberries.

But in greater abundance was the sights and sounds of nature. The rustle of leaves in the breeze as the hardwood trees danced with the wind. The angry squawking of a Blue Jay or barking of a squirrel as we disturbed their progress in whatever act we had caught them. The silver flash of a fish swimming to get away from our splashes as we made our way upstream. The occasional streak of shiny-black as a snake would quickly slither towards a rocky outcrop and away from the invaders.

Eventually, we would reach our island. A sandbar had long ago formed in the middle of the creek and settled long enough for a few trees and grasses to grow, which helped hold the island in place in spite of the occasional flood or springtime thaws that brought extra amounts of water down from the hills to greet this mid-stream obstacle.

Once we arrived, we would pitch our tent, an Army shelter half set up as a lean-to. We'd arrange the dry rocks into a circle for our campfire and gather dry wood to prepare for the night. And we would explore.

We had a spring that was dug out a little, along the side of the creek where we could go to get fresh drinking water. This worked as a perfect place to fill our canteens unless in one's haste to fill the dipping cup, one muddied the waters. I learned quickly to work more slowly and save time in the long run for such activities.

Mealtime was so great in many ways. Bacon, potatoes, eggs, corned beef hash, an onion, and some salt and pepper, all cooked in a cast iron skillet over a campfire of logs and sticks. And when we were finished, the creek was the sink in which we washed our dishes, the sandy bottom was the soap and scouring pad. Sometimes we would eat Army C-Rations. The round paddies of vanilla and chocolate fudge being my all-time favorite candy bar in the world. Then, and even now as I reminisce about candy no longer available, anywhere.

At night the woods didn't sleep. It changed a bit, with the dark shadows and sounds of night-time creatures who, in the darkness, transformed from normal animals into beings of monsters imagined. The small screech owl made a night-call sounding all the same as an insane criminal, or some type of Bigfoot hunting for campers to terrorize. The dusk to dawn symphony of crickets, croakers, and cicadas was non-stop and got louder each time you forgot to listen. And the birds didn't all sleep at night. Their cries were sounded from all corners and at times would make your neck hairs raise up. And with all of that, the sleep I would sleep was the best I could get, knowing that on our island, with my dad, I had nothing to worry about. With the exception perhaps, of muddying our spring water again.

We found many things while exploring those woods. I remember an old graveyard we found that had long been forgotten and was covered with vines, bushes, and shrubs. We found sections of a frontier time, corduroy road that had been laid by settlers moving westward into the Ohio Valley region, perhaps on their way to a future Fort Boonesborough.

Though, not everything always went according to plan. We were chased out one trip by a big storm. The lightening and especially the risk of rising water gave us plenty of reason to pack up and depart early. Also, one time we were given the opportunity to walk out much further than we had planned after the battery went dead in the car during our camping trip. This was before everyone carried a phone, and the only option was to walk. For me, the walk out of the area was an adventure. For my dad, not so much. Having been told not to tell mom about the dead battery, I managed to keep our secret until about half way through dinner. We survived, nonetheless.

Again, the island was a special island. This was the island that during my last visit as a child, my dad marked the occasion in a special way by fastening one of our forks to a tree to leave behind. Soon after our campout, he was going to Vietnam and had made a promise to come back to the tree and retrieve our fork in a year. This was a promise made and a promise kept. Eventually. That story is also an extraordinary story.

Father's Day Story covering June 1968 and August 1991

In 1998, The Gazette, a local Colorado Springs newspaper had asked for some stories about people's fathers for consideration as part of a Father's Day special to be run in the paper. I wrote the following story about an event that my dad and I had shared many years earlier, just before he had left for a tour in Vietnam. The paper selected my story and I was interviewed for a short write up, and had my story printed in the paper. I sent my dad a copy of the newspaper article and am including the story here as it was originally written for the article.

The Fork in the Tree

I'm a fourth-generation soldier and the story I'd like to pass on is the hub around which my relationship with my father has taken several revolutions. My father meant everything to me as I was growing up and I was always content in the fact that I wanted to be just like him. It was after his trip to fight in Vietnam, that our relationship made its most dramatic change. For a long time after his return, and subsequent divorce from my mother, I felt he had abandoned me to be with his new wife. Our relationship began healing in bits and pieces as I entered the Armed Forces and also became married and finally became a father myself. We really became closer than ever before, when I returned from my own war in the oil fields of Kuwait and we were able to share the fact that we had both been partakers of the ancient ritual of combat. This is when the climax to the story I will share took place.

It was June 1968 and my father had just buried his own father while explaining to me, an eleven-year-old boy, why people must die and how it was not always fair, but that it always happened someday. He also held the burden that he must tell me he was leaving to go and help stop the spread of communism in South East Asia. Even at that young age I had heard about, and believed, the domino theory would prevail, if left unchecked.

Before he left, we went on a camping trip to one of our favorite spots in the deep woods of Kentucky, along Otter Creek, on a tiny sand bar island. I told my Dad, I was afraid when he left he may never come back. Meanwhile, my Dad was still in mourning for his own father, who did leave and would never come home. My Dad took one of our metal forks we always brought camping and pounded it into a tree, promising to come back and return to the island and retrieve the fork. He would not die in Vietnam he told me, and I believed him.

When he returned, many variables entered the picture over the next two decades and we never went back to our island. I had several dreams over the years of returning and finding the fork.

At some point during the long wait, I even thought we had actually returned and seen the fork, but we never had. It was just my mind, perhaps my hopes, playing a trick on me. Twenty-three years after we had placed the fork into the tree trunk, my Dad, my own son Joshua and I, took a hike back into the woods to look for our fork.

Up and down the creek we walked, ever looking for something from our memory to come into view, but time and weather have a way of changing things. Our island was no longer in the creek. Had it washed away, or was it simply a part of the normal creek bank now? We had looked all afternoon and needed to get back towards the car and had all but given up when my Dad told me, "Son, we have come full circle." The island had moved out of the flowing stream due to a change in the way the water was running down the valley floor. Our island was now just part of the bank, like so many other miles of woods lining the creek banks. But, in a thicket, barely visible, my dad had spotted the now rusty fork protruding from the trunk of a now, much larger tree. What was really unique was the fact that a limb, or sprout of the first tree, had grown up and encompassed, actually grown around part of the fork's handle, holding it locked in place to the spot where it had been pounded into the trunk of the tree so long before. As my father so eloquently told it, "as God wrapped his arm around us and kept us safe all those years, so had the tree wrapped its own arm around the fork, keeping it secure until we were able to return and fulfil the promise that had been made on the island so long ago."

As I had just fulfilled a similar promise to my own son and daughter, to safely return from the Gulf War, my own young son had no problem understanding how important this event was in his father and grandfather's lives. We decided to leave the fork in the tree and return home. Later that year, as a Christmas present, my Dad presented me with the section of tree to which the fork had been pounded into, the securing limb that had grown around the fork, and of course, the fork, all encased in a cabinet he had made for me as a very special gift.

Inscribed on a plaque he had attached to the case was the following fulfillment of a promise made and a promise kept.

In June, 1968, a father had just buried his father and was preparing to go to war. He and his young son went on a final camping trip. That night, on a little sandpit island, on a deep run of Otter Creek, the son told his dad that he was afraid he might never come back. His father had left him only a month before and would truly never return. He pounded a fork into the tree that they were sitting under. He promised his son that he would return and bring him back to get their fork.

Twenty-three years and many incidents beyond their control passed. God placed a fatherly arm around the son and his dad. Just as soundly, he grew a protecting limb around the fork. Passing time and unchanging love has finally closed the door on a promise made so long ago. Here son, is our fork. I told you I would come back. I love you, Dad. August 1991

Vietnam and Everything Changed, Athens, Ohio, 1968 – 1969

Things were really starting to get interesting on Ft Knox. There were hundreds of UH-1H "Huey" helicopters flying over the neighborhoods and troops training all the time as Vietnam raged and the amounts of troops kept increasing in both numbers in-country and numbers of those killed in action.

Meanwhile, for my family, things were pretty normal. My dad had taken me out to a really cool house on post that was where a person holding the job of a military Game Warden would live. Apparently, he had talked to someone about getting a special assignment to work outside his normal career field, which was Armor, and specifically, working in the Armor School at the time. He must have been pretty sure that he would get the special duty assignment and seemed to be quite excited about not only the job, but also moving to live in the house. At the time, we lived in typical military housing for junior NCO level soldiers, which consisted of a two-story apartment in a building with seven or eight other apartments. Living in a stand-alone house with only our family in it would be a new experience, and it was located on a big piece of property near the woods.

Then suddenly, our plans and hopes were changed. My dad received orders to go to Vietnam.

I remember when he told us. I acted like I was going to cry, but in a strange way I was excited to think that he would be able to go fight the enemy and safeguard our country. Although I was young, I still kept an eye and an ear on the news about Vietnam. Of course, we talked about it at school and many of my friends were worried that their dads were either going, or they were already there. Now, in a weird way, I was feeling proud that my dad would go do his time there also. I never considered that he would be hurt or killed, at least not until he actually left.

I soon discovered that we were not going to stay at Ft Knox while he went to Vietnam for a year. I don't know if we even had the option, but we were going to go live in my dad and mom's hometown of Athens, Ohio. I was happy about that because my grandparents also lived there and I knew I'd be spending time with them.

Once we moved and before my dad left for Vietnam, we set up our new rental house. It was a big house with a huge yard. I had never lived anywhere so fancy before and I was super excited. My dad and I cut a long pole down in the woods and fashioned it into a flag pole. We put it in front of the house and I raised the flag every day while he was gone.

During his tour, our nation put the first man on the moon. Many things changed during his time spent fighting in a war. The US was in a state of unrest as young people held protests and riots across the country against the Vietnam War. There were also race riots and black people fought against the police in many major cities. For me, it was a time when many deep feelings of patriotism, but also darker feelings of disdain for what I viewed as anti-American behavior began to develop. I didn't like hippies, I didn't like black power protesters, I didn't like college student who protested against the war. I also didn't like a lot of things I started to see happen in my own home. My mom and dad were growing apart while he was in Vietnam and I saw it happening, just as I saw what felt like the whole country growing apart. And then, my dad returned.

My dad came back and landed in the middle of a situation already out of control and headed for a crash. He and my mom quickly separated and got a divorce. I ended up staying with my mom. My dad went to his next assignment with a new wife who I had known recently as the teenaged, girl next door. It was a hard time for everyone, but we moved on.

Our nice house on the hill was traded for an old and small rental near the railroad tracks on the outskirts of an older part of the town. We lived there a year or so as my mom found a job and got her feet on the ground. We eventually moved to a small mobile home that she had bought and placed on her mom's land a few miles outside of Athens, on the property where she had grown up.

My dad had meanwhile spent a tour of duty as an instructor for the ROTC program at Kent State University in Ohio, where he lived through the protests, riots, and shooting of the four students by the Ohio National Guardsmen during protests on 04 May, 1970. After this, he received a more traditional military assignment.

The Riots of Athens Ohio

My seventh-grade year happened during a tumultuous time in the United States. During 1968, Martin Luther King Jr, an African-American advocate for Civil Rights was assassinated during his support of a protest focusing on Civil Rights. Later that year, Bobby Kennedy, the brother of the assassinated President John F Kennedy was also assassinated after he became a Presidential candidate himself. Both 1968 and 1969 saw the United States in chaos across the nation as riots occurred in major cities over the ongoing war in Vietnam, protests against the military draft, and protests over the discrimination against black Americans.

It was in the aftermath of one of these riots that my normal walk to school created a memory that was far from normal. The city of Athens is home to Ohio University and in 1968, the college students at the university had overran the streets and created enough damage and trouble during the nightly protests, which ultimately became crime-filled riots, that the Governor had called in the Ohio National Guard to assist the local police department in maintaining order and protecting the city and the people.

I had to walk across the main part of town to get from my house to the school building. This morning as I neared the city-center, I saw soldiers dressed in combat uniforms standing at every-other parking meter along Court Street, the main street of the town. Each soldier was holding a rifle, the standard M-14 rifle, and each rifle was tipped by a very sharp bayonet. This was quite an unusual sight, to say the least.

I, being the son of a soldier and a very pro-military minded young man, felt a kinship with the soldiers as I knew they were there to do battle against the same people that I also considered foes of the American way of life, the rioters of the night before.

As I neared the lines of Guardsmen along both sides of the street, I paused to speak with a few of them. At some point, unbeknownst to me, I had touched something that was still covered with the residue of a tear gas canister that had been set off the prior night.

Once I had finished talking to the soldiers as I passed by on my way to school, I settled down in the auditorium awaiting the first bell, the one telling us to go to our homerooms and get started. With the unrest and general atmosphere of anger being felt across the country by people from both sides of every issue, we no longer loitered outside playing or hanging around, waiting for school to begin. The auditorium was a safe space for us during those days of uncertainty. It wasn't long before I reached up to rub my eye and accidentally and immediately discovered just what tear gas was all about.

Soon I was feeling better after washing my eye in the bathroom sink and getting a few paper towels wet with my seemingly, never-ending stream of tears. It didn't hurt, but the waterworks just didn't quit for a while. I thought it was cool in a way, but not that cool.

The riots and chaos continued throughout the school year, but eventually things calmed down and returned to what was considered normal. Still, to me Seventh Grade seemed normal enough and I looked forward to the end of the year for all the normal reasons, not really thinking about the politics and other particulars that the people on the news were always upset over.

As I attended Eighth Grade, my mom moved us out of the city and into the country. I managed to befriend a neighbor who soon began to treat me like one of his own two sons. I started learning to ride horses, work with cattle, harvest hay, and do some hunting and fishing.

It wasn't long until I was in High School and started seeing life so much differently each and every year. My focus was not on the world, not on the nation, not even on the state. My focus was all about a few friends, fast cars, and having fun.

About the time I turned sixteen and started driving, I also started working. A car and a paycheck led to dating. And dating quickly led to Linda. As I started paying more attention to Linda,

I rapidly quit paying attention to almost anything else. Especially politics, school, and even work. My life changed and continued to keep changing in a big way.

A Girl Named Linda Elkins

Linda, or mom, or grandma, wasn't always my wife. She was once a girl I met after we had agreed to go on a double-date with a pair of mutual friends. Linda went to school, a different high school than mine, with Elaine, the girlfriend of my best friend, Butch.

Butch had recently lost his license for a while after getting caught drag racing by the cops. Butch wanted to take Elaine to the Homecoming Dance at her school, Alexander High School in Albany, Ohio. He had asked me if I would drive his car so he could go and pick up Elaine to take her to the dance. I was told that Linda Elkins had seen me around the school at games and dances, and said that she'd go to the upcoming dance with me if I wanted to go. I thought that was a great idea, so I said yes.

A few days before we were to go to the Homecoming Dance, I was watching TV at home, Bonanza, I believe. I heard someone pull into the driveway and then someone knocked on the door. It was Linda. Her friends had brought her to my house to meet me before we went on the Homecoming date. I invited her in and we talked a while and planned for our first evening of what would turn out to be, a life-long adventure. Now I suppose that this specific visit was partially inspired by the fact that over the last several weeks, Butch and Elaine had tried to set me and Linda up a couple of other times, but it never worked out. I blew off the first attempt at a date because I decided to go to a football game with my buddies instead. The next time I was supposed to meet her, she decided to do something else and failed to show up. Perhaps she was teaching me a lesson, perhaps she thought I wouldn't show. The next date, the Homecoming Dance, was to be our 3rd attempt. Three's a charm, so the saying goes. She must have wanted to make sure that I was serious about going, since after all, it was her Homecoming Dance. Either making sure that I was serious, or simple curiosity must have produced the drive-by interview. Before texts, iPhones, and email, people actually had a lot more face-to-face conversations.

After our impromptu meeting, we said goodbye and she left. While she had been visiting, I noticed that she was wearing a quite-large Future Farmers of America (FFA) jacket. I also noticed that it had a name on the front. Joe Caldwell. I couldn't help but wonder just who this Joe Caldwell was and why Linda was wearing his jacket, which by-the-way, was about four-sizes bigger than anything I would wear. Was this a boyfriend? An ex? Was she wearing it on purpose to let me see it?

All of that was something I thought about, but chose to ignore because I really wanted to go on the date after meeting her and I didn't care if some unknown person named Joe liked it or not.

Linda's High School's Homecoming Dance night came and all went well. I drove Butch's hot rod, a souped-up 1963 Ford Galaxy 500 for the event. We picked both of the girls up at Elaine's house and drove on to the high school. Linda was very pretty and was wearing a dark blue velvet mini-dress, which was normal length for the time back in 1972, normal being short. With heels, she was almost taller than I, and would be for the next couple of years until I became taller as I got closer to 18. But as I mentioned, she was very pretty and I quickly decided that I was interested in dating more than once even if she was a little taller than I was. That was if she would be willing to go out again with me.

Apparently, the feeling was mutual and a few more dates in September and October of 1972 turned quickly into going steady. I gave her my High School ring to wear a couple of weeks before Halloween. We became an item from then on, and I never looked back. Butch and Elaine lasted a few months and then they broke up, but Linda and I were "going steady" our Junior and Senior years of High School and decided we wanted to get married sometime around the summer of 1974, soon after we had both graduated. We had graduated the same year, but from different High Schools; she from Alexander High School of Albany and I from Athens High School.

Oh, and as for Joe and the FFA jacket? He was Linda's brother-in-law, married to her sister Deb. She didn't even think about the jacket when she came out to visit that night, and we have laughed about it later, more than once.

Dating and Other Things

Once Linda and I started dating, we let the rest of the world sort of just go away. We enjoyed each other's company and simply put, couldn't get enough of each other. We were each other's real first romantic interest. Pretty soon it was obvious to everyone else as well.

Linda was raised by her grandparents, her dad's parents. She and her sister Deb had gone to live with Arvel and Cuma Elkins after their mom and dad got divorced while they were very young. They were the daughters that Linda's grandparents never had, having five sons, including one who had died of Scarlet Fever as a baby. Linda was an outdoors girl and spent much of her time growing up following her grandfather, or Papaw as she called him, around the farm and fields as he went out to work in the garden or hayfields. Deb was more of a dolls and house sort of girl, and stayed closer to home, although they would both sometimes go outside and play all day, as was acceptable back when times were a bit more worry-free.

Linda was 17 when I started dating her and I was sweet-16. Her parents, grandparents actually, didn't think I was particularly sweet. In fact, for quite a while, her grandma didn't care for me much at all. Which was fair in a way because my mom didn't care much for Linda at first. She believed an "Albany girl" was somehow going to get me in trouble. At that time, I drove a very loud car and thanks to my job as a mechanic at the service station in town, I generally smelled of grease, oil, and gasoline. I was a Hot Rod guy and during the time I was dating Linda I drove a 1965 Chevelle, a '67 Mustang, and a '68 GTO Pontiac. All loud and I drove them hard and fast. This did not go unnoticed by her Grandparents, and especially her grandma, which was also who she called mom.

I was not the ideal material for their daughter as far as she was concerned. I did get along quite well with her Papaw though. We grew close over the time I dated Linda and he became something of a second dad to me. Eventually I also won over her grandma, although it took a bit of work. I also think that I helped win her over by introducing her to pizza. Once in a while I would bring a pizza mix over and make it for everyone. She and Paps liked pizza once they tried it and I reaped the benefits of getting in their good graces through my pizza-making ability.

Looking back at myself as a 16 to 18-year-old, I can easily see why I would not be the first choice for someone's daughter as a date, let alone as a steady boyfriend. I was rough around the edges, to say the least. I was pretty used to running hard with little to no restraint, and I tended to follow a path that inevitably led to troubles at many of the turns. Nothing really bad, and nothing the law was concerned about, besides fast driving. I just pushed the envelope a bit too much at times. I liked Linda, I liked my fast cars, and I liked having Linda with me, in my fast car, going places and doing things that were not pointed in college or much of a career-oriented goal.

One of the first times Linda's mom met me, was at a car wreck I'd been part of. My friend and I had skipped school and decided to pick Linda up when she got out of classes at her own school. He drove, since it was his car, and after I talked to Linda for a while, we took her home and dropped her off. As we left her place, my friend got a bit too deep in the throttle and as we crested a small hill on a curve, he lost control and crashed less than a quarter-mile from Linda's house. She later said that she actually heard the crunch of his car as it rolled three or four times and ended up down an embankment, on the edge of a small creek's bank.

I had been thrown out of the car at some point during the crash. No one wore seatbelts back then, and I had landed in the creek, unconscious, with only my feet showing above the water. My buddy Steve told me later that as he crawled out of the car, he saw my boots sticking out of the water and ran up the bank, to the bridge. I had apparently raised myself out of the water and stood up momentarily, only to fall back beneath the water and beneath the thin layer of ice that covered the small creek. Thankfully, Steve was OK and he was about to get me out of this situation that his failure to stay on the roadway had put me in in the first place. Just then, someone had stopped to see what had happened and to help. My buddy Steve told them to go call an ambulance and then he jumped off the bridge and into the water to pull me out.

I don't remember any of this. Once I did wake up, I was lying beside the road, wet, and covered by a blanket. Linda was there and was quite upset. A few other people were there. And Linda's grandma was there and she was quite angry. Who could blame her? She was giving me the Devil for having had Linda in the car minutes before the crash. Someone in the crowd that had gathered told her to leave me alone and then offered me a cigarette. I said that I didn't smoke and that generated a positive response from several of the people in the crowd. It's funny what sticks out in one's mind after something like this happens. Eventually an ambulance, in the form of a long station wagon that doubled as a hearse, showed up and drove me to the hospital emergency room. After a few stiches in my head and a few more in my knee, I was fine and anxious to get back to Linda's house to let her know that I was OK. Of course, by then, my mom was there. My friend's mom was there also. Everyone was angry. And that's sort of the big starting point for my and Linda's relationship.

Clandestine Communications

Back in 1972, the computer was not a household appliance. In fact, people didn't even have smart phones. Unless the telephone was attached to the wall of your house with wires, you didn't even have a telephone to talk to someone. On the good side, there were less distractions to be concerned about while driving. Texting while driving was not even a thing yet.

Every Sunday morning, Linda would go to church with her Papaw and Mamaw. They would leave the farm, drive across the county all the way to a place called Bearwallow Church of Christ, near Coolville, Ohio. That is, until I started dating Linda and we decided Sundays were better spent together discussing things of more importance to the world as we saw it.

This was not a favorable option according to her grandparents though. We had come up with a plan, however, that on occasion, Linda could get away with staying home for an assortment of reasons, and then I could come visit while everyone else was at church for most of the day. As I mentioned earlier, there were no cell phones back then and I couldn't just drive out to Linda's house and risk getting too close before her grandparents left for church. As I also mentioned, I liked fast cars and my fast cars also meant very loud cars.

Necessity being the mother of invention, we came up with a plan. I would drive to a place on the road where I could see her place, but not be seen myself. As her grandparents left for church, in the opposite direction, Linda would go throw a rug over the fence near a large tree. From my viewing spot along the road, I could see if the rug had been placed over the fence and then I would know if it was safe to drive up to the house. There are several ways that this could have gone wrong, but we never got caught.

Later in life, as I was learning certain aspects of unconventional warfare and its related clandestine communication techniques, I always thought about Linda's safe signal of the rug over the fence, as I created communication plans for scenarios requiring two-way communications in a denied area.

Riding Horses

When I started dating Linda, I found out that she loved horses. As a child, she used to ride on the back of the horse as her Papaw was plowing the garden. She even has memories of taking an unexpected nap on the back of the horse as it pulled the plow around the field and gently rocked her back and forth into sleep. She had a buckskin horse named Dusty when I met her. I would help her get hay out of the barn loft and feed her horse as we talked about school or our plans for the weekend. We also enjoyed riding and went on many trips out and about, following cow trails through the pasture fields or along the gravel roads that crisscrossed the countryside surrounding the property.

Being that she only had one horse, when we went riding we had to double up on the horse. I remember riding in the snow during the winter and riding double had the extra benefit of helping us to stay warm. But we also rode during the summer and enjoyed the time our rides would give us to get away from the farm and all that kept us busy. I remember stopping and having a picnic in the woods, or just hanging out while the horse ate some grass or rested in the shade of the trees. Papaw sold the horse once we got married and left to go live in Germany, but during my break in the service after returning to live in Albany, we got into horses in a big way.

Papaw deeded us 5-acres of land from a nice large pasture field and we bought a new-to-us 14x60 foot trailer. We had the trailer pulled to the property and once it was set up and all the electric and plumbing was put in place, I was able to get Linda a horse again. I got another buckskin horse and its name was Sandy. I got this horse from a friend of ours through a trade. I traded a .357 magnum Ruger Blackhawk six-shooter and holster for the horse. I had bought the pistol while we were stationed at Ft Carson, CO. Many times since, I wished I had kept the pistol and figured out some other way to have gotten the horse, but at the time it seemed like the best thing to do.

We soon traded or sold the buckskin and we ended up with Birdie, an Appaloosa with a really pretty white spotted rump. Then we got a grey Quarter Horse named Camey. Eventually we sold Birdie the Appaloosa and bought a young two-year-old bay mare. We called the young horse Star due to a star-shaped white spot on her forehead. I broke Star to not only ride, but also to work in harness.

All of the horses we had prior to getting Star had been registered American Quarter Horse or Appaloosa horses. Star was what we called field grade, which meant she was not papered. She was obviously of Quarter Horse blood though. She was also one of the best horses I've ever seen. She was a very calm natured and steady thinking trail horse, and could pull a log in harness and never spook or buck. Lori learned to ride and could control Star by herself (under close supervision) at the tender age of just 4 or 5 years old. We even took Lori and Star to a few horse shows and she got some ribbons and trophies during her short cowgirl career.

Of course, the horses came with a lot of baggage and all of it was expensive. Between the saddles, bridles, blankets, and feed, there was veterinary care for routine checks, and occasionally an emergency visit. Another expense was the horse trailer and truck that could pull it. The license plate on the front of the truck said it all; Go Broke, Buy a Horse. But we had a good time when we went riding, and I learned how to take care of horses, to include becoming a pretty good Ferrier. Our horse time was a good time. Expensive, but fun.

The Great Horse Race

Jim and Pam Ferris were some really good friends of mine and Linda's. I had bought our first horse from Jim. Actually, traded is more accurate. We would go on trail rides once in a while and usually we had pretty normal rides. One ride was a bit on the extra-ordinary side, or at least a part of the ride was special. We trailered to a trailhead in a state park and the four of us rode through some very nice woods and forest areas. It had rained recently and parts of the trail were pretty muddy and some of it was covered with water. Some of the water holes were pretty large and almost resembled small ponds.

Eventually, we ended up at a really cool saloon. It was made from rough-cut lumber and resembled a saloon you would see in an old western movie. It obviously catered to people riding horses in the area and there were even hitching posts set up in front of the place, next to normal parking places for cars and trucks. We tied the horses up and walked into the place through the old-style swinging doors. We could have been walking in from a 19th Century cattle drive and were dressed the part. Boots, hats, and even spurs that made me think of a spaghetti western or some Clint Eastwood movie as we walked across the floor and headed to the bar. It was summer and we had been riding quite a while. A cold beer was just what the doctor had ordered and one beer turned into another and perhaps another.

Realizing that we had to ride back the same distance we had come, we finally headed back out to the horses and got them ready to go back to the trailer. They had all had water and were rested and ready. A well-broke and happy horse enjoys trail rides as much, or more than the people riding. My horse, her registered name being Ding-a-Ling Camey, was more than ready to get back on the road. Before I bought her, she had been a race horse and had the high-strung personality to go with the background. She didn't just walk, she pranced everywhere and at the touch of my heels against her flanks, she would fly like the wind.

Jim was riding a horse that he also considered fast and he had done pretty well in different horse events like barrel racing and the like. Needless to say, between the horses being rested and heading back home, as well as me and Jim having had just about enough beer to make us feel bulletproof, it wasn't long before we decided to have a horse race.

My mount could smell it before I even started getting her lined up and she was ready to go. When we took off, my horse's first jump must have been 20 feet down the trail and we were out front and running hard. Jim and his horse were trying, but never did catch us. Camey had a way of moving into a gear one notch higher than full-out run when you turned her loose, and she was hitting on all cylinders this afternoon. I was a good rider, but as she raced along the trail I was a little concerned by her eagerness to throw caution to the wind, but I had enjoyed enough beer that I didn't do anything about that bit of concern I was feeling.

Now I'm sure we only raced about a mile at the most, but it seemed longer than that as we went around each curve and twist as we followed the trail through the woods. Suddenly as we came around one curve in the trail, the ground to our front disappeared and was replaced by one of those small ponds I remembered passing by on the ride in the other direction earlier that morning.

I know that I had about 100 ideas go through my mind in a second or two as I looked for an exit. I thought about bailing off, but figured that the fall would probably kill me. I thought about trying to hit the brakes, but there was no time to stop. A crazy, fast horse and a lack of sense (beer-induced) had put me in a tight spot. I decided the best thing to do was to trust Camey, give her plenty of rein, and hang on. I'm not, nor ever was a mind-reader, but I'm pretty sure Camey was thinking "hold your beer and watch this" as she seemed to accelerate before launching herself up and out over the water.

I would have loved to have seen what Jim saw from behind as we made like an act in a wild west show. Camey, with me stuck like glue to her back, had to have made a leap across that water hole that was 50 feet or more. There were big trees on both sides of the water and nothing but jeep tracks beneath. Jeep tracks and mud covered by a foot of water or more. Anyway, she cleared the water and stuck the landing without missing a beat. She was fine, but I was finished. I started wheeling her in and slowed back down to the speed of sound and then got her stopped. I don't know who was breathing the hardest, but we were both still sound and healthy. Jim came riding up, eventually followed by the girls, and he told me that he figured he was going to watch me, and maybe the horse, both die in a terrible crash. He said he had never seen anything like that before. I was proud of the way it had turned out, but all my buzz from the beer and all my bravado from before the race had drained away when we launched rocket-style over that big hole of water. We went on back to the trailers and trucks at a normal pace and I brushed the mud off of Camey and rubbed her down before loading her in the trailer for the trip home. I would make sure she got a good supper that night. I'm glad, still today, that we came in first, but I also am thankful that I had the beer and she had the water while we were at the saloon. It might have been a story with a different ending if I'd have had whiskey and had ordered beer for my horses.

Regretfully, the next time I turned Camey loose in similar fashion, the girth strap on the saddle broke and I was the one that launched through the air. I came off her back, saddle and all, at about Mach 2 and headed toward the downhill side of the hill we were running across. As I fell to the left, my right shoulder was the first thing that hit the ground. That fall landed me in the hospital, after waiting way too long to go to the ER (thanks machismo) for a badly broken shoulder that needed surgery to put it back together. Being good horse people, I was back on Camey and we were trail riding before I got my arm out of the sling and the metal rods out of my shoulder. I might have gone broke for a while by owning horses, but at least I didn't break my neck. Close to a broken neck a time or two, but I survived.

After we had both graduated from high school we started to think about the future. At some point we just sort of started to assume that we would get married and didn't go through a fancy proposal or engagement, we just became more serious and started to make some plans.

I was living in a small rental cabin in Albany by this time, and was working as a mechanic at the ARCO gas station just down the street. Besides doing mechanic work, I also worked with tires, to include large truck and even tractor tires. I was making enough money to keep gas in my 1968 GTO, but not enough to really consider getting married and supporting a family.

I had always planned on joining the Army as I was growing up and I reminded Linda about this every once in a while. Now that I had finished high school and had no intentions of going to college, I started to get serious about the Army idea. I talked to a recruiter and took the entry test and had a green light to join. Linda was not familiar with the military life at all, as no one in her immediate family had ever been in the military, let alone make a career of the military.

We discussed the Army as an option that would allow us to get married soon. I knew if I didn't have a good job that we would be unable to get married for a while, and would have to wait until I could become better situated. Linda agreed that joining the Army sounded like a good idea.

My dad was in the Army and at this time was stationed at Fort Knox, KY. We decided to take a trip there to visit and see what the Army looked like, at least from a closer perspective than I would get from the recruiter. All seemed well and I entered the Army in March, 1975, a little less than a year after I graduated. Basic Training was about seven-weeks long and I found myself in a green uniform early in March as a member of the Delta Demons. My unit was D Company, 13th Battalion, 4th Training Brigade. My dad, coincidently, was a senior NCO in the training unit's higher HQ. During my training, sometimes my dad would show up and check in on me. My Drill Sergeant would always make sure these visits provided me with some extra attention after my dad would depart in the form of push-ups. They obviously liked my dad, one of their senior NCOs, and wanted to make sure that his son received a good deal of training.

In fact, my first day in the unit had me setting up my wall locker and making my bunk, following specific instructions we had been provided from the Drill Sergeants. I was in a large bay with about 30 other new soldiers and we were working hard trying to meet the timed requirement we had been given. About the time that I was almost finished, one of my Drill Sergeants came up to me and I snapped into Parade Rest. He looked at my gear, looked at the name on my fatigue shirt, and then looked me in the eyes and asked "Is your old man an E7 who works in Cherry Hall?" Apparently, the HQ building where my dad worked and was the NCOIC (the Sergeant in charge), was named Cherry Hall. Thankfully I knew this from my visits before I swore in with my Oath of Enlistment and started training. I told the Drill Sergeant that yes, he was my dad. With that bit of information my Drill Sergeant told me to pack my stuff back up because I was moving to a Squad Leader's room. His reason was "With SFC Easterling as your dad, you should be used to this" and I immediately became Squad Leader and remained one for the rest of Basic Training. While it came with a lot of extra responsibility and stress, it also resulted in a quick promotion upon graduating to PV2 (2nd rank, one above a Private E1) and my "mosquito wing" rank getting sewn to my green dress uniform. I was so proud of that small yellow strip of cloth, perhaps more than anything I'd seen up to that point.

Now Basic Training was an event all by itself, but I, well, Linda and I, decided to make it even more special. With the help of my dad and his 2nd wife (Sally), Linda had been preparing for our wedding, which was to take place right there in the training area at the Cavalry Chapel, the same place we trainees marched to for Church Services each Sunday morning.

Linda had hoped for a nice wedding in Albany, OH. One with her family and our friends in attendance. But the Army had decided that my schedule would not permit the timing to be what we were hoping for, so we did what we would do a lot while in the military, we allowed the situation to dictate.

My graduation from Basic Training was on a Friday afternoon. I was released from the company area on a Saturday morning to go get married. We were married that afternoon with one of my fellow trainees as a Best Man, and then we went bowling with my dad and his wife Sally. Linda and I spent our honeymoon night in a small hotel in Radcliff, KY, just up the road from the Ft Knox entrance. The next morning, I had to return to my company and load a bus to be transported to Fort Sill, OK for my advanced training in my new job as an Artilleryman. Linda would drive my car back to Albany from Ft Knox and wait a couple of weeks until I was able to see her again.

Oklahoma to Germany

After about a week at Ft Sill, I was able to get a weekend pass and I flew back to Ohio. Linda and I packed up a few things and tossed them into the trunk of my GTO and we headed southwest to the land of tornados and heat like we had never seen before.

We were to be in Lawton, OK for about 8-weeks while I finished my Advanced Individual Training (AIT) to become a Fire Direction Control Specialist, or as the Army called is, a 13E, FDC Specialist. I had wanted to be a Tanker, someone in Armor who worked with the M60 Tanks. However, when I got to Ft Knox (the Home of Armor at the time), I was told that there were no openings in Armor and if I wanted that, I would have to wait. Meanwhile I could join and go with the Artillery, which had big tracked vehicles with cannons attached to them. They even look a lot like tanks, I was told.

Sometimes however, one can be too smart for one's own good.

I joined and went to Basic as a potential Artilleryman. In the Artillery there are a couple of jobs for the new guys. A 13B, or what's called a Gun Bunny or Cannon Cocker, worked on the guns that "look like tanks" and they actually get to fire the cannons. For a potential Artilleryman who scores above average on his entry tests to get into the Army, a different job awaited. That's where I landed. I didn't get to work on the tank-like guns, I just got to tell them where to shoot. I was never satisfied with this job, even though I excelled and was promoted quickly. This dissatisfaction from the start caused several major events in the future, but ultimately, they all led me towards bigger and brighter results.

But, back in Lawton, OK, Linda and I moved into a small rental, which was actually half a house, and we shared the house with the owner, an older, and very nice woman who lived alone.

While I was not on Post, learning about FDC, Linda and I were enjoying our honeymoon trip away from everything we had ever known. The weekends were filled with adventure as we explored the area and saw Indian Reservation land, buffalo, and even some nearby tornado activity. We also went swimming one weekend and got so sunburned by the hot Oklahoma sun that we thought we might never recover. But we did. All too soon, it was time to pack up and drive back to Ohio.

I was going to Germany to serve in the Field Artillery as a member of the 8th Infantry Division. I had only a few days to get ready and once more, I left my new bride behind and traveled alone.

While in transit as I moved from temporary holding facilities between Frankfurt, to Bad Kreuznach, to Baumholder, I got a tattoo. It was an eagle clutching a banner with Linda written across the banner. It would be a surprise for Linda once we were able to get back together.

After a few weeks in Germany, I was able to find an apartment in a small village named Nohfeldon. We would spend the next two years or so, living in the upstairs apartment of the Hans Wagner family house. At least Linda would.

Germany was an eye-opener for us both. I quickly found out that my job in the Field Artillery had a huge emphasis on the word, field. I was gone, out in the field, more than I was home. I was also gone for extended periods of time that stretched into weeks. Linda had to do a lot of soul searching during this time as she suddenly found herself alone in a foreign country with no means of transportation, no way to speak to anyone, and no knowledge of how to make the situation any better.

For me, it was not as bad, but it was a huge change from what I had ever experienced before. I was only 18 when we got married, when I joined the Army, and as I started to learn how to do my new job in a foreign land. Oh yes, and how to act as a married man with a wife that was dependent on me in ways she had never had to be dependent on anyone before, due to the circumstances of our current living arrangements.

I was a good soldier and quickly moved up the ranks. This was the time period when the Army had just ended the Vietnam War and started the new VOLAR, or Volunteer Army. Many of the guys I worked with had originally been drafted. Many had served in Vietnam. Suddenly, at only 19, I made my Sergeant rank. As an E5 or Buck Sergeant, I now outranked and was put in charge of quite a few men who had fought in Vietnam and had been in the Army for years longer than me. It was a challenge at times, but I enjoyed most of what I was doing. Linda meanwhile, was getting tired of being alone in Germany.

While I was home, or when I came home most every afternoon after work, Linda and I had a great time. We travelled the countryside, as much as possible with our beat-up crap car we had. We knew a couple from Tennessee that we'd met and they became our best friends. We went many places with them and had a ton of fun. As long as I was not gone, Linda and I had a good time.

Still, after about a year and a half, Linda decided, as did I, that it would be better if she went back "to the world" and stayed with her family until I got assigned back to the States. We didn't have any children yet, so we thought that this was the best thing to do for all concerned. She would go home and help her family, go back to work at her old job, and save some money. I could move back to the barracks, also save some money, and focus on my job, which had gotten quite hectic with the upcoming spring and summer exercises being planned. In August 1977, I departed Germany and was assigned to Ft Carson, CO, where we spent the next two years and grew our family by one, our daughter Lori. Fort Carson, however, was not the US Army, Europe.

Reforger 1976, Baumholder, Germany

Reforger is a military acronym that stands for Return of Forces to Germany. While I was stationed at Baumholder, Germany as a member of the 8th Infantry Division's forces, we participated in the September 1976 Reforger exercise and went against the 101st Airborne Infantry Division from Ft Campbell, KY. As a Germany-based unit, we were going to use tactics and strategies common with the former preeminent adversary, the USSR or Soviet Union. Our Commanding General had given a speech at the beginning of the planning period and stated that we would be on the 101st Airborne like a Chicken on a June Bug. Soon, bumper stickers were everywhere with our unit's insignia over the Screaming Eagle of the 101st's unit patch with the logo "Like a Chicken on a June Bug" and the morale was as high as I'd ever seen it.

Reforger was a different type of training event. During the 1970s, the exercise area was the entirety of Germany, not just designated military training locations. Our unit deployed across our actual assigned areas of Germany, which happened to be within the Fulda Gap region. We would maneuver through villages with streets barely wide enough to allow our tracked and armored vehicles to pass between houses and buildings. There was always a massive bill turned in to the US Military after the exercise was completed for damages to German property. It was a joke that if you ran over a chicken, the US government would have to pay for the chicken, all the eggs it would ever have, and all the chickens that would have ever hatched.

As the exercise began, I was detailed to be my Commander's designated driver instead of my normal job. The Commander's vehicle was a standard Army Jeep, the M151, a quarter-ton utility vehicle. I was told that the Battery Commander wanted someone who could not only drive safely, but could read a map, operate the Command Net radio, and follow the play of the exercise as it unfolded.

I soon discovered why this all was important, and why he wanted me as his driver. Every time he got into the Jeep for a movement to do a recon or lead the main body to our next location, he went to sleep. This was apparently the only time he was able to sleep. When not driving, he was being kept busy during any period we were stopped with the unit activities.

I also would assist my actual FDC section whenever I got the chance during periods where we would be in one position for more than a couple of hours. I got a chance to really see the countryside during the exercise and I even got to meet a lot of Germans as we set up our positions in field or woods located just outside towns and villages in our areas. Some of the older Germans had watched American units near their towns before, during WWII. Likewise, they had also seen the Nazis as the war raged between their own military and the Allies who defeated Hitler and the Nazi military.

Some of the younger people would come out to the positions where we set up and visit or even bring food, and sometimes beer. After the first several days or so, it was determined that even while limiting our tactics and weapon capabilities to those of the USSR, we were putting a big hurt on the 101st Airborne units. We were given the authorization for the second half of the exercise to start using our true US capabilities and we began to beat them even harder. The 101st Airborne Division was taken to task quite severely by our units and we enjoyed the praise we received from our own Commanders. I ended up getting promoted to Sergeant (E5) in less than two-years after joining the Army, and the quick promotion was in part due to my performance during Reforger 76.

In the mid 1970s, the Army was still switching from a draft-based Army to a Volunteer Army. Many of the soldiers in my unit that I soon out-ranked, had been drafted or had served in Vietnam. As a 20-year-old brand new Buck SGT, I had my authority tested more than once. However, my personality and demonstrated abilities as a soldier helped me achieve a widely accepted "Green Tab" or NCO leadership role among the majority of soldiers holding all ranks. Reforger was a great opportunity for me to learn and apply all the skills I'd been learning since Basic Training.

Good Times in Germany, 1975 – 1977

My first tour in Germany was a good one in a lot of ways. It was tough at times, and it was really difficult on Linda at first. We got married and went straight to Germany, and after having never been away from her home before, she was not only away from home for the first time, and by herself a lot, she was living in a foreign land where she didn't know her way around and couldn't speak the language. We basically leaped before we looked. We really hadn't thought this one through.

We lived in a small German town quite a few miles from the US facilities where I was stationed. At first, I didn't have a car and I hitched a ride from the one other US neighbor who also lived in the town, or I hitch-hiked to get to work. Then, I'd hitch rides to come home at the end of the day.

As soon as possible, I bought a clunker car. With it, we were finally able to go shopping and even drive around sightseeing during some of the days I was home. Being in a field unit, that is, a unit that was always in the field, Linda spent a lot of time in our apartment reading books, magazines, and painting a lot of paint-by-number pictures I'd pick up at the PX. There was no American TV either; only the Armed Forces Radio, which wasn't always playing the songs we wanted to hear. And there was just one station on the radio speaking English. Those were the good old days.

Our apartment was actually part of the second floor of a German family's house in a town named Nohfeldon. We had a small kitchen, and a living room and bedroom area that was divided by a curtain. We shared the upstairs bathroom with the family's washing machine. Also, the oldest child, a daughter of about 11 or 12, had a bedroom on the 2nd floor located off the common hallway we shared. But, Linda and I were both young and newly married. Being together whenever possible, under any conditions, was better than being apart.

During the first few days after Linda arrived, we met what were to become our best friends while we lived there. Dewayne and Rita were a couple from Tennessee who stopped to gave us a ride as we were hitching our way home from the commissary one day. We immediately hit it off and started running around together.

One of our favorite things to do was visit a nightclub named the Disco Tanz in the small town of Birkenfeld. Birkenfeld was right in the middle of a triangle created by Idar-Oberstein where our friends lived, Baumholder where I was stationed, and Nohfeldon where we lived. It was about 12 to 15 miles from any of these points to the town of Birkenfeld which made it a good place to meet. The Disco Tanz was also where I was introduced to Jägermeister. This German liqueur was a favorite of the bartender who worked at the club and every time I would order drinks, I'd always be asked "and Jägermeister for me?" with a big smile. I always felt welcome, even though as Americans, we were a small minority of the regulars who were mostly locals, young German adults.

We also attended the Oktoberfest in Munich with our friends from Tennessee. We drove to Munich and set up a tent in one of the campgrounds within walking distance to the Fest. We were young enough to enjoy the rides and still manage to drink a lot of the German beer that came in liter-sized mugs in any of the dozens of huge beer hall tents and buildings set up on the festival grounds. At some point we managed to find our way back to the tent and get some sleep before we had to get up, pack up, and start driving back north towards home. Slightly hung over and no doubt smelling a bit like a horse, but honestly happy.

We were all so poor and didn't even know it. Driving junky cars and living pay check to pay check never really interfered with the amount of fun we could have on any given day. A picnic basket and a grassy field along the road provided us with ample excuses to get outside and simply enjoy life.

And then, there was the movie theater. About once a month, we'd go to one of the US movie theaters on the main base at Baumholder, or hit the one located on a small outlying station nearer our apartment. It was always a great time to see an American movie and be carried for a couple of hours back across the ocean to what we all called “the world” which was the USA.

Eating was also a treat in Germany. Once in a while, we’d save our marks, the old pre-European Union era German money called a Deutsche Mark, and go to a Gasthaus for some fine German cuisine. More often than not, instead of a Gasthaus, we would go to the German’s version of a fast food joint. The Schnell Imbiss stand was usually a small trailer setting along the roadway where you could buy Brats or Bratwurst, Currywurst, and Schnitzel sandwiches. And of course, Pommes Frites, the German French Fries.

The roadside chow in Germany was some of the best food in the world. Served in thin paper wrappers, and usually saturated in oil. The smell and taste of this delicacy of poor-folk food was, and continues to be wonderous.

Living in Germany as a teen-aged, low-paid, American soldier, with a wife, was tough and it took a lot of internal fortitude for us both, but we still had some of the best times of our life during those two years. To paraphrase what’s been said before, we were both soldiers, and we were young.

From Germany to Colorado, 1977 – 1979

About April or May of 1977, Linda decided that she wanted to go visit home and go back to work at her old job to help save some money for our upcoming move. We didn't have any kids yet and I was extremely busy with an uptick in the amount of time I was spending downrange. So, Linda went to Ohio and I stayed in Germany until August.

In August of 1977, I left Germany with 30-days of leave or vacation, before I had to report to my next duty station at Ft Carson, CO. I flew into Ohio and was picked up at the Columbus Airport by Linda, along with her sister Debbie and bro-in-law Joe.

We had a great time visiting with family and friends and I enjoyed a small welcome-home party. A lot had changed in two-years, but a lot had stayed the same. In fact, I noticed that for the people who had stayed home, nothing much at all had changed. Meanwhile, in spite of all the hard times we'd had in Germany, we also had a lot of new and exciting adventures. We had seen and done things that everyone at home had only heard about. It was a bit of an eye-opener and a confirmation of what we had hoped, joining the Army had been the right choice in the first place.

I'd been home a little over a week when we got the news that Elvis Presley had died. That was a big deal for many people as Elvis was a generational icon. He was only 42 and died of a heart attack while sitting on the toilet in his bedroom at his Graceland mansion in Memphis, TN. I remember thinking that he was so rich and famous, but he was dead and I was still alive. I was the lucky one.

While I was home, we picked our brand-new car up from a dealership in Athens. We had ordered the car from the PX in Germany and had it shipped from the factory in Michigan to Athens. Soon, we were driving to Colorado with a backseat full of our necessities. We arrived at Colorado Springs and drove around looking for an apartment complex. We found one and it looked decent enough, so we were soon in a one-bedroom apartment on the west side of Colorado Springs, about 15 miles north of Ft Carson. The next day, I reported in and started finding out what the "Stateside Army" was all about. I quickly found out, I didn't like it.

The Army at Ft Carson was a lot different than it had been in Germany. The soldiers seemed to belong to a different Army and the overall attitude was one I had only observed in a small group of soldiers while in Germany. There was a big problem with drugs and a bigger problem with racism at Ft Carson.

Initially, I attempted to manage what I saw as a leadership problem the same way I handled myself in Germany as a young and newly promoted Sergeant, an NCO. I quickly discovered that being "hard-core" was not appreciated in this new Army I'd found myself in, by either the rank and file nor the senior leaders.

I quickly adjusted myself to the environment and did the best I could with what I had available. I did pretty good as the Section Chief of my Fire Direction Control team until a senior NCO got assigned and took over the position. He was probably one of the sorriest NCOs I had ever seen and he only made a bad situation worse. Again, I adjusted to the environment and decided that the Army was no longer for me. I was now counting the days until my time was finished and then, I was going to go back to Ohio and get a job at the local coal mines where my brother-in-law Joe was working.

Meanwhile, Linda and I decided that we should have a baby while I was still in the Army and had the medical coverage we enjoyed. Almost as soon as we knew we were medically and physically ready to start trying, Linda got pregnant. By mid-September, 1978, we had our baby girl Lori. Five-months later, I turned in my gear and we said goodbye to the Army life.

A couple of people, my Battery Commander and one of the Section Chiefs of the M109 Howitzer gun sections had tried to convince me to reenlist and give it another try. Go back to Germany, get promoted to Staff Sergeant, I was even offered a Gun Crew and the opportunity to work on the gun-line instead of the FDC if I would only reup. I wasn't having any of it though, and told them I would never think about staying in or coming back.

Little did I know that four-years later, I would not only come back, I'd agree to come back as a Private and start all over again. I found out that besides missing the good parts of being a Soldier, I also missed the security of the medical coverage for myself and my family. Also, the guaranteed, albeit low, paycheck each month. Mostly though, I missed being a Soldier and that's something you couldn't find as a civilian. At least, not back in the early 1980s.

Death Valley Days, Ft Irwin, CA, 1978

Even during my time at Ft Carson from mid-77 until early 79, I still had a few good times as a Soldier. Operation Red Flag took my unit from Ft Carson, CO to Ft Irwin, CA and downrange into the Mohave Desert where war games had been played for decades. There were still tank tracks made by General George Patton's tanks as they trained for war in Europe. The entire region had been used as an impact area at one time or another and shrapnel or fired bullets could easily be found just beneath the sand that covered everything.

At the time I deployed to Ft Irwin, I was the Chief Computer for the Fire Direction Control section of a 155mm Self-propelled howitzer battery. We had six M109 Howitzers in our unit from Alpha Battery, 1st Battalion, 29th Field Artillery. We were known as the Alpha Animals and instead of wearing the standard green ball cap when we were downrange, we wore Vietnam-era floppy jungle hats with desert camouflage. It was a small thing, but provided a huge return in esprit-de-corps.

After our several weeks of being out in the field and conducting war games, we ended up with a couple of days with nothing on the schedule before we redeployed to Ft Carson. Somehow, we convinced the Commander to allow the Alpha Animals to take one of the vehicles and travel cross-country over the desert, to Death Valley for an overnight visit and return.

Using a map and compass, we navigated the open desert and forged some flash-flood areas that showed up after some unexpected rain. As we reached the far bank of one wide stretch of water that had been about waste deep as I crossed it with one of the other Sergeants, he asked if I thought the truck could make it across. I replied that I didn't think it could, as I waved the truck forward and into the water. The truck got in pretty deep, but the old WWII era Jeep pickup truck just kept chugging along and was soon on dry land once again.

Within several hours, we were in Death Valley proper, the actual National Park. We went to several sights, to include the Bad Water Basin. This is the lowest elevation in the US and at 282 feet below sea level, what water you find there is not something you want to drink. Once inside the actual park we followed the road to Furnace Creek and stayed overnight at one of the campgrounds.

While there we visited with some of the other campers as the night turned into the next morning. Soon a Park Ranger showed up and asked us what we were doing and whether we had explosives in our vehicle. Apparently during the night, one of the members of our group had made the comment that we were there to clear out some rocks that had covered one of the roads during a landslide caused by the latest rain. Someone told the authorities that some soldiers with explosives in their truck were staying in the campsite, and that set off some alarm bells for the Park Rangers.

Once we assured the Ranger that we were just enjoying the sights like everyone else, we started back across the desert. We managed to find our way back in time to get the truck cleaned up and turned back in, although it was a bit worse for the wear and tear we had put on it. But, aside from a few funny looks, we got everything cleared and we were ready to climb into the C-130 for the flight back to Ft Carson.

Alpha's Animals had survived to cause trouble another day.

Goodbye Army, Hello Ohio, Hello Army & Germany Again

As a recap, I left Germany after my 1st tour with the 8th Infantry Division, stopped in Albany, Ohio for a few weeks of leave, and to pick up Linda before we headed off to Ft. Carson, CO. At Ft Carson I was assigned to another Field Artillery unit, only instead of being the 8-inch self-propelled howitzers, the unit's fire power now was the 155mm self-propelled howitzer.

I quickly found out that the Army at Ft Carson was a lot different than the Army in Europe. I soon began to become disillusioned with what I was seeing and decided that when my enlistment was up, I would get out of the Army and go back to Ohio.

During our assignment at Ft Carson, we had been married a bit over three years and we were anxious to have a child. Our daughter Lori was born at the Ft Carson Army Hospital. It was a large, old complex comprised of wooden buildings of the WWII vintage, connected by sidewalks covered with small rooftops and a few enclosed passageways. In fact, Lori was born in one building where the nurse showed her to me just before wrapping her in blankets and taking her outside and over to another building to get her to the nursery. The old hospital has long-since been replaced by a state-of-the-art hospital.

I had been talking to my brother-in-law Joe, the same Joe who owned the jacket Linda was wearing the first time we met, and he thought he would be able to get me a job at the Southern Ohio Coal Company where he worked. I thought being a coal miner would be a great job, and it paid a great wage. We were looking forward to going home and being civilians again and around the middle of February, 1979, we pointed the AMX east and headed for Ohio.

This turned out to be the same time that a lot of negative attention was being paid to the coal industry by the politicians in DC. The Sulphur found in much of the coal from the region was too high for the EPA to ignore. As the concerns over pollution continued to be addressed by various regulations and requirements to make coal more environmentally friendly, the bottom of the coal industry started to drop out. Once I was actually back in Ohio and was looking to get hired by the coal company, they were no longer hiring.

As a Veteran, I was able to attend an education program created for the region's primary employer, that being the coal industry. I decided that attending the school would probably help me get hired. After a couple of months learning how to work as an underground coal miner, I was officially designated as a United Mine Workers Association-certified Miner's Apprentice.

The certificate and a buck would get me a cup of coffee in most places, but it didn't help get me a job. Coal mines all through the region were talking about lay-offs and closing down mines. No one was hiring. That's when Joe had me talk to his brother Bruce, who worked at a place called the Austin Powder Company. This was a dynamite plant and they were hiring. Soon after, I was working as a "powder monkey" making dynamite. It didn't pay as good as being a coal miner, but it paid well compared to about anything else in the area. And they had medical insurance for the family, so I was once again a gainfully-employed, happy camper.

Red Hat Hunt, Coal Miner Dream, Powhattan Point, OH

I'd worked hard for my Underground Miner Apprentice Certification. I not only spent a lot of time in a practice mine, I got to tour a couple real ones in hopes of actually going to work in one. I successfully completed an 80-hour or two-week course to become an official Underground Miner Apprentice Certified, Red Hat wannabee. The course was taught at a location just down the Ohio River from Wheeling, West Virginia, nearby Powhattan Point, OH.

The classroom was next to a large structure made of concrete blocks with a roof and no windows. On the inside, it looked like an underground coal mine with entryways, crosscuts, working faces, and even coal. When you went into the mock-up mine, it was very realistic, very dark, and at times, very dusty.

As a new miner, you would have to wear a red hardhat for your first year on the job. This was so you could be easily identified as someone still learning what your job was and let everyone else know that you needed some extra attention. Learning the ropes came fast and furious, even in a school environment. While in the mock-up mine, we ran the real equipment and it could hurt you just as bad as any of the same things underground in a real mine.

I have a lot of good memories about the school and the quick, but long-lasting education I received. Of course, my family on both sides, as well as (my wife) Linda's family were multi-generational coal mining families. I have little doubt that had I got hired after attending the Miner's Apprentice program, I would never have left the miner's life behind, at least not until I retired or got killed. I felt comfortable with the idea of having a miner's career the same way I did about having a soldier's career. Nothing really to think about, it just seemed natural. Although, there is a patriotic satisfaction that would not have been fulfilled by working as a coal miner. Perhaps, truth be told, I was born a soldier. But during this time in my life, I wanted to be a coal miner in the worst way.

The underground vehicles were my favorite. The Buggy followed the Continuous Miner and was the workhorse, at least from my point of view. A Buggy would pull up to the hopper side of the Continuous Miner and the onboard conveyor and the Buggy's arms would transfer the coal from the Miner to the Buggy until it was full. Then the Buggy would head to the beltline, which carried the coal, on a conveyor belt system, all the way outside to the coal yard. The Buggy would transfer each load to the moving beltline, drive back to the Miner and get another load. That was the sequence and it was repeated time and time again until the end of shift.

The Continuous Miner had a big rotating head with diamond-tipped bits, or teeth, on a round steel drum. The drum would spin as it moved up and down as the operator moved it back and forth across the working face, the place where the coal was being dug out of the coal seam.

The drum would pile the coal up in front of the Miner and two arms would pull it onto a conveyor that brought it to the other side of the machine and into a small hopper.

Since you couldn't do a three-point turn in the tunnels, (main, return, or cuts), the operator's area, called a Kitchen, was designed to be driven from both directions. The steering wheel was on the inboard side of the buggy and the operator would switch his seat from one side to the other, depending on which way he was going. Steering one way with your left hand and the other way with your right.

Another vehicle was the Scoop Car or Loader. This was kind of like a big dust pan on wheels. The operator, driving from the Kitchen, had a big scoop bucket on the front and would scoop up whatever and carry it wherever. A general-purpose type vehicle that carried coal, debris, tools, or supplies.

The Roof Bolter was a cross between a floor jack and a drill. It would drill the steel roof bolts, about 3-feet long, into the roof, or what was called the top of the mine. Another piece of equipment was called a Mantrip. The school's mock-up mine didn't have one of these, but one of the mines we visited in WV had one. Like an underground bus, it carried you and the crew beneath the ground and to the worksite, sometimes a couple miles down and miles along entryways or crosscuts.

However, there's a unique feature about all of these vehicles which I've yet to discuss. These vehicles are only about three feet tall. You operate them from a laying or semi-setting position because these machines are operating in spaces that sometimes are just inches higher than the top of the machine.

Also, the machines are electric powered. Most are not battery powered, but are connected to huge electric cables that spool in or out, depending on which direction the machine is going. The other end of the cable is attached to huge power centers located near the working areas. These are connected to large electric cables running outside. Basically, high-power electric cables are everywhere underground. And often, so is water. Add to that the fact that sometimes the cables get cut, ripped loose, or pinched, and you can see one more dangerous aspect of working underground. We were taught to always step on the cable as you crossed over one, if possible, just in case the cable popped up as it was being wound in by the equipment to which it was attached. Having your leg picked up by a cable was thought better than having your crotch smacked by a cable when it popped upwards.

Inside the mines, almost everything was covered in white lime dust. The top, the sides or ribs as the side walls were called, even the bottom was covered with lime dust to keep the coal dust from floating and creating an explosion hazard. The only really dusty places were the working face, the place where the coal was being removed, loaded, and transferred to the beltline for its trip outside and off to a final destination.

Speaking of explosions, electrical connectors and cables, and other things that could injure a miner only highlights the need for the large amount of safety and basic medical training we also received. A couple of unique safety items were the Self-Rescuer and the Methane Detector. Methane, a poison and explosive gas, naturally occurs in coal. The utmost care must be practiced to monitor it and ensure that it doesn't build up to dangerous levels. The detector, like the old canary in a coal mine, would always be used to test the face prior to any shift starting operations. As the Miner was running, sparks would ignite small pockets of methane gas in the coal seam and small flashes and fireballs could often be seen as the drum ripped the coal loose.

The Self-Rescuer was a final-option device each Coal Miner wore on his equipment belt. In the event of a fire, the carbon monoxide (CO) would quickly build up and if the ventilation failed, the Self-Rescuer could save someone's life. Like a small catalytic converter on an automobile, the device would turn CO into carbon dioxide or CO2. However, the higher the level of CO, the hotter the device would get, with the possibility of making it so hot it would burn the mouth of the user. They would only last an hour or two, so you needed to find a way out quickly if possible.

The last characteristic that stuck in my mind, is the ventilation process. The air flow is what kept everyone alive. Way outside are huge fans. They blow air into the mine. To control the airflow, the miners used heavy, mostly yellow-colored, plastic coated curtains, like large sheets of canvas. These curtains would direct the fresh air to places where the miners were working, run it across the really dusty locations to clear the air, and then run it back outside through the returns. In theory, and unless something got screwed up, in practice, this kept the air fresh for the miners and sent all the bad air, coal dust particles, gases, and anything else you didn't want to breath or smell, to the outside. Obviously, the return airway was not a place you wanted to spend much time.

But, with the new EPA regulations and the demand for the types of coal found along the Ohio River, I wasn't able to get a job with one of the coal companies nearby where we lived.

Not too long after I completed this school and found that I was likely not going to get hired any time soon, I got another opportunity for employment.

Linda's grandfather and grandmother had raised both her and her sister. When I first left the Army, we lived with them for a short while. Pretty soon, her grandfather gave us a five-acre section of his farm and we bought a used, but nice, 14-foot by 65-foot mobile home to set on the hilltop of our new homestead and call it our castle. We really enjoyed living there and we felt like we were doing great. Linda was working as well, and Lori was growing like a weed. I had remained in the Army Reserves, although I never had to attend any training.

About that time, I started working for the Powder Plant. Then, after about three years at the dynamite plant, I got laid off in a big reduction of forces. I was told that the plant may never open again, or that it would never operate at full strength if it did reopen. But before I leave Austin Powder for greener fields, let me share some memories about the time I spent as a Powder Monkey.

Austin Powder Plant, Dynamite Maker, McArthur, Ohio 1979 – 1982

Powder Monkey

From near the end of 1979 until the summer of 1982, I worked for the Austin Powder Company and made dynamite and various other explosives. When I say made it, I mean that I and those I worked with actually created the end product by starting from making nitro glycerin and mixing it into the other ingredients, placing that mixture into paper or plastic wrappers, packing those wrapped tubes or bags into boxes and sending them down the plant's production system to the shipping department.

The laborers, of which I was one that held the job title of Utility Man, were all Teamster Union members. Our foremen were the non-union company hires and as such, were not just the bosses, but in many instances were also the targets of a lot of juvenile behavior. We, an unusual and quite rowdy workforce, mixed working inside an explosion waiting to happen with other activities to make the environment more enjoyable and entertaining. At least from our perspective.

The Complex

The plant was actually a lot of individual buildings spread across and around a series of hills in an area that still contained most of the trees of the forest in which it was built. The trees, along with large berms of earth and timbers, produced barriers that would help block the blast of an accidental detonation from any adjacent buildings. The layout was designed to utilize gravity as the primary engine to move the product, the explosive materials being produced, from the top of the mountain through the complex facilities and finally on downward to the roads and railway at the bottom of the hill where the administrative and non-explosive production work was accomplished. Why was the gravity aspect important? It had everything to do with safety. The nitro and the raw explosives were so volatile that using gravity kept potential detonators, like a spark or friction from a vehicle with an engine from igniting an explosion. Even a spark from a mule's steel shoe against a flint rock or metal tram rail could be bad. Gravity could safely provide the effort to move the liquid nitro glycerin or loaded trams of raw explosives downhill to places it was packed, secured, and shipped.

The buildings in the complex that didn't contain any type of explosive product, the Dope House, machine shops, carpenter shops, admin offices, the shower or locker building and the like, were mostly made of brick and quite old. The buildings where the explosives were manufactured were all made of wood and looked like small white houses. Each one was built inside a space that had been dug out of the side of the hill, which with the barricades would tend to direct any unintended explosions upward. The buildings were heated by steam which was produced by the boiler building down the hill in the admin area, and the boiling water that the steam pipes produced was collected in barrels at each building to use during weekly scrubbings of all workplaces each Friday afternoon. All electric points were sealed and met explosive environment requirements. Smokers had to leave the hill and go to the parking lots to burn one during breaks or lunch, and no smoking materials were allowed to be brought up the hill. Almost everyone chewed some type of tobacco and every corner of every building was always brown with tobacco spit stains. Each building also had a loud siren with a switch placed at each door so that if an emergency took place the people working could hit the siren and warn everyone else as they ran for their lives on their way out. That's if they could see it coming.

Explosions were not frequent, but they did take place. There were a couple of places where you could see old tram track, like narrow-gauge railroad track, still wrapped around large trees, having ended up there during an explosion in one of the buildings. Needless to say, personnel who were in a building when it exploded were almost completely gone afterward and there were only a few small fragments ever recovered from the explosions that I knew about. All the explosions I ever heard about had happened on the Gelatin Line or the Fuse Line. I spent most of my time on the Powder Line. The Fuse Line was a part of Austin Powder, but was under a different union and on a more distant site not connected to the powder and gelatin lines.

High on the mountaintop sat the NG Factory. This is where the nitro glycerin was made. It was produced in big tanks made of lead and each batch was measured, even though it was liquid, by the pound. The lead was important in a couple of ways. Nitro is heavy and like mercury can sink into almost anything. The lead is dense enough to prevent the nitro from fouling or contaminating the lead. Also, lead lowers the spark probability, which could be catastrophic for a tank holding thousands of pounds of liquid nitro glycerin. A batch or tank of NG weighed about 12,000 pounds. The NG was white like milk due to some sort of antifreeze or stabilizer, and was much heavier than water. Obviously, it was also highly explosive, but not as touchy as it's portrayed in most movies or cartoons.

The tanks would start out with a few inches of water on the bottom to cushion the load of NG as it poured into the tank through a heavy rubber hose. The NG would settle through the water and then the water would float on top of the NG, protecting it as it sat in the tank. From the NG Factory, the NG would flow through a series of rubber hoses held in wooden troughs to another building called a Halfway House. This was the cut-out between the Powder Line and the NG Factory. From the Halfway House, the NG would be transferred to the Oil House. NG was referred to as oil as it was sort of consistent with how oil felt. At the Oil House, the NG would be stored in the same type of lead tanks, and as needed, the NG or oil would be drained out of the large tank into a smaller rail-mounted tank attached to a scale. Various types of explosive mixes required different amounts of NG. The tank could hold up to 250 pounds of NG. From the Oil House, the movable tank would traverse an enclosed pathway to the next stop, the Mix House. As the name implies, the NG would be transferred via another set of the heavy rubber hoses, still using gravity as all this distance was on a continuous downward slope of the mountain, into another tank attached to the large wooden mix bowl. Here, the oil would be mixed into the other non-explosive ingredients. After this point, there was no more raw or liquid NG to handle.

The Mix House itself had two mix bowls. They were wooden bowls that resembled a wooden livestock trough if it was shaped in a circle about ten feet in diameter. Inside the bowl were two large wooden wheels about eight inches wide and five feet in diameter. They were suspended from a post extending up through the ceiling and hung about one inch above the bottom of the bowl. There were two small plows attached to a cross beam just in front of the wheels to push the ingredients being mixed into the center of the bowl, which the wheel would then run over and push back towards the sides. This action, coupled with the slow draining of the NG tank into the mix, which also rotated around the bowl on the frame holding the wooden wheels and plows is how the bowl worked. Once the NG was mixed into the Dope, the wheels would be stopped and the mix was hand-shoveled out of the bowl with wooden or plastic shovels into carts by the Mix House operators and helper.

From the Mix House, the numbered mixes would be transferred via two wheelbarrow styled carts per mix, to the appropriate Pack House. This again was done using gravity and a tram, or wheeled cart that traveled on small gauge railroad tracks. A Trucker would ride the tram and apply the brake while transporting the mixes down the tracks. Each Trucker had a mule, a real 4-legged mule, not a tractor, to use to pull the tram back up the tracks to the Mix House to get another load. There were two Truckers on the Powder Line, an upper and lower Trucker. The upper Trucker handled moving the mixes to the Pack Houses from the Mix House and the lower Trucker took the product from the Pack Houses to the Case House.

At the Pack House, the mixes, which usually ranged from 1000 to 1200 pounds each, would be used to fill paper or cardboard tubes and started to resemble the normal dynamite sticks seen in the movies. They ranged from the standard six or eight-inch-long by 1 ¼ inch in diameter sticks, up to ones that were 24 inches long and sometimes up to six inches in diameter. Sometimes the mix would be placed into plastic bags resembling sandwich bags, depending on what type of explosive was being produced. Once more, the Trucker would then transport these items with his tram further down the hill to the Case House, where the product was placed into boxes for movement to the shipping yard and on to the end customer.

The complex consisted of two lines, the powder, or dynamite line I've just covered, and a similar set up for the gelatin line. The two lines were on opposite sides of the mountain and both converged at the base.

The area at the base is where the administration buildings were located, as well as the building that produced the various mixtures of inert ingredients to be pulled up the hill to the Mix Houses. These inert ingredients were pulled up the hills on trams that were pulled by a small diesel locomotive. The mixtures were put together in a large brick building called the Dope House and the Dope was loaded into metal barrels to be delivered up the hill to the Mix Houses. At the base of the hill there was also a shower house, as all employees were to shower after each shift. Special clothing had to be worn while up the hill to avoid sparks and contamination.

Besides the obvious danger of working with explosives, being exposed to the nitro glycerin each day had its own medical drawbacks. Each week, all employees had to receive a blood pressure check to make sure that the NG was not causing problems in that area. It took most people a couple of weeks to get used to being around the NG, and almost everyone suffered headaches and nausea, sometimes severe, during the first few days of work. Once your body was used to it, you could start feeling withdrawal symptoms by the end of a weekend and felt better after taking a deep breath of NG infused air once back on the hill.

The equipment for all the manufacturing of the product, from the NG Factory on down the line to the Case House looked like items you would see in pictures from the 1800s. In fact, most of it was exactly like it was designed back during that time period. Everything was made from wood, to include the shovels to move the product from the mix bowls into the carts. Then, the carts had to be emptied with more wooden shovels at the Pack House and loaded up onto a hopper attached to the packing machines. The work was hard and the men working were truly a special or different breed of men not found in such quantity in many other career fields, other than ones with obvious and real dangers, such as coal miners or soldiers. Almost all the employees were Veterans of the military and that was by design.

The People

Trust was paramount within the personnel who worked the lines. Literally, lives were dependent on how well everyone followed the safety regulations and standard rules. My foreman was a WWII veteran who had been a POW after the Nazis captured him in Germany. He was a fair, but quite gruff boss and he put up with little in the way of nonsense. His name was Paul Molihan, and we all called him Mollie. He was also a big Ham Radio operator and his radio handle was Powder Monkey. He figured we were all Powder Monkeys and it seemed like a good name to me.

In writing this story, I started to recall the men with whom I worked the closest and have included their names mainly because I can still remember them after over 40 years. I never felt like I belonged, in the sense that I was somewhat of an outsider while many of them had grown up with each other, or were from very similar backgrounds. I usually would feel most distant when we were eating lunch and relaxing with normal or routine conversation. Often, especially during the first few months I worked there, I'd make mention of something that was going on in the news from around the world. I could ask a question like, "did you see what was happening with the government of Egypt?" I almost always was met with a couple of glances and silence. Then someone else would say "I got me a new coon dog" or something else about hunting or fishing, or drinking, and the conversation would continue in that direction. It would be wrong to say that these guys were not smart, but they were focused on their lives and the small circles in which they existed.

The upper Trucker was my best friend at the job site and his name was Jack Bartoe, one of three brothers who worked on the same Powder Line. His older brother Larry was the Oil Pusher, and his younger brother Roger worked in the Case House.

Jack had gotten burned across his back while in Vietnam and used to tell the story once in a while. He was on the perimeter of a Fire Base and one of the other soldiers had picked up a flare that had failed to burn the night before after being shot out of a mortar. He had apparently scraped the powder from inside the flare and asked Jack for his lighter to set it on fire. When it caught fire, it exploded in a fireball and burnt Jack's back. He spent several months in hospitals recovering.

Few stayed for long in the Mix House as it was a very physical job and most didn't care for the way the exposure to the raw NG worked on one's body. Ernie Black and Steve Durham were Mix House operators, as was John Shepherd. John, also a Vietnam Vet, eventually got hurt at the powder plant while working to break in a new mule. Sometimes we would break-in new mules after we had finished with the daily standard at the Mix House. The mule he was breaking into the job pulled a tram off the tracks after becoming frightened by another vehicle and pinned John between the tram and a porch post on a Pack House. He recovered and went back to work, but it crushed his face and head pretty bad and he was never quite the same as before.

There were also a couple of guys I worked with a lot, the other (late shift) Utilityman Keith Fee, and the guy who worked in the ALX building (scrap, waste, and bad explosives), Roger Exline.

Of no real importance, but something that was true, there were no black men working at the Powder Plant, and there was a story of only one who had ever worked there before, but quit after the first day when he found out it was not a talcum powder plant.

I don't know if that's true, but again, the entire population of the Powder Plant, Union and Company, was made up of a rough breed of man and a couple of women on the night shift who worked on the Gelatin Line.

I was hired to fill a Pack House Helper job before I switched jobs to the Utility Man position. There were two packing machines in each Pack House. Each Pack House had two operators and one helper. The helper had to keep both hoppers full for the packing machine operators or the quota of packed product couldn't be met by the end of shift. This requirement kept the helper very busy, especially if the packing machine operators were fast and the machines didn't break down. After just a few days, I was told to come in on the early shift, an hour earlier, and start working as a Mix House Helper. The Mix House had two mix bowls and two operators. I would be the one helper. My job was to shovel half of the mixed dynamite mixes out of each mix bowl. That meant going from one, straight to the other, usually without any break in-between.

It was a hard job, but I got good at it and stayed on the early shift even after bidding for and winning a Utilityman position as my assigned job. As a Utilityman, I had to learn all the jobs on the entire line. However, since the Mix House was a very demanding job and required the crew to be in an hour earlier than everyone else, it was hard to keep someone there full-time. As a Utility Man, I worked almost every day at the Mix House.

I soon learned not only all of the jobs on the line, but I also learned how to "push oil" which meant I could run the Oil House and transfer NG from the tanks between the NG Factory and the Oil House, and bring the NG from the Oil House to the Mix House.

The early crew consisted of me, two others from the Mix House, the Trucker (and his mule), the Maintenance Man, the Oil Pusher, and of course, the Foreman. We were all Vets and a few had served in Vietnam. As events played out, in general, it was the early crew against everyone else when it came to anything besides actual work. This was a rowdy bunch and where possible, the rowdiness came out.

Nitroglycerin Shower

One day we were mixing ditching dynamite. This was a heavy NG mixture which was 500 pounds of wood pulp and 500 pounds of NG. The wood pulp was strained and screened and very dusty until the NG was added and then once it was mixed, it had to be sent to the Pack House and packed up quickly to keep the NG from seeping out of the mixed wood pulp. Since the NG tanks on the mix bowls only held up to 250 pounds, the NG had to be mixed into the wood pulp from two loads, one tank full at a time.

When things are running smoothly in the Mix House, the three workers all help each other get the job done quickly. As two are working one bowl, the 3rd is setting the other bowl up for the next mix while waiting on the first bowl to finish mixing. No one stood around and waited on something to finish, and we all worked to get done as quickly as possible. Once we had mixed our assigned standard from the Foreman's instructions, we could clean up and be finished working until quitting time.

This day, there were several of these mixes on the order and every mix of the ditching dynamite had to have two times of running the tank dry into the mix. The Oil Pusher would usually bring a tank of NG to the Mix House and after the Mix House man hooked up the hose, the Oil Pusher would drop the 1st load of NG into the tank. The Mix House man (me) would start the bowl and begin draining the 1st tank. Once it emptied, the Mix House man would stop the bowl and hook the hoses back up and keep working until the Oil Pusher returned with the 2nd load of NG. Then the Mix House man would go back to the tank and make sure the hose stayed in place as the 2nd load was emptied into the tank.

This particular day, the Oil Pusher dropped the 1st load into the tank and didn't inform the Mix House crew it had been delivered. When he came back with the 2nd tank, the hose was still hooked up from when he dropped in the 1st load. This time he signaled the Mix House that he was ready to drop the load of NG. I thought, after having done this many times, that he was there with the 1st load. I walked to the mix bowl and held the hose to make sure it was secure and told him to drop the load.

Just after he started the NG flowing into the tank, it overfilled and when the lid came off the NG sprayed all over me, the mix bowl and the floor where I was standing. I was now soaked in nitro glycerin and standing in a puddle of it, surrounded by several thousand pounds of high explosives.

The Oil Pusher stopped the flow of NG and disconnected from the Mix House. I remained motionless as he went back to the Oil House and called the Maintenance Man and the Foreman. The Maintenance Man brought paper towels and alcohol to wipe me and the floor and the equipment off. As I had each shoe cleaned off, I picked my foot up and removed the shoe and sock. Then I had to remove my clothing and wipe down with the alcohol. Once things had been cleaned and I could move, I went out of the Mix House and got new work clothes and boots from the Foreman who picked some up for me at supply down front in the administration area.

Once we made sure that there was no hidden NG anywhere, we restarted the mix bowl and finished mixing the NG into the pulp. Since this was now a bad mix, it had to go be worked into our "junk" explosives that was used mainly by strip mines. This was what we did with any mix that ended up made incorrectly. Likewise, all of the spilled explosives that got swept up from the floors of the pack houses during each shift became what was called ALX and was sold in tubes that weighed 25 pounds each. It was not guaranteed to be of any standard and could be 25 pounds of high explosives, or it could be 25 pounds of pure wood pulp. Surprise explosives.

No one got into any trouble over the spill, at least none of the workers did. The Foreman had to cut several mixes from the daily standard since the Mix House got behind in its total mixes. I got to experience being soaked with NG and we all got a slight raise in our adrenaline levels when the NG shot out of the tank and sprayed everything with some of the most dangerous explosive material to be found.

In the military, I'm sure someone or several people would have gotten written up and maybe even had to lose some rank. We, however, were Teamsters and in the union, and we were almost untouchable because of it.

A Grease Job and the New Foreman

There were a lot of pranks played between the guys on the Powder Line. Nothing that was dangerous, but sometimes they would get a bit out of hand. I was one who would occasionally go a little too far. Good old grease, the black gooey grease you find on axles, or the white clean grease you find on special equipment tended to be the heart of many pranks.

Powder clothes were white. The t-shirts, the pants, the gloves. Powder clothes were removed and replaced with clean clothes before the smokers would go off the hill to get a nicotine fix. Obviously, they were the targets of opportunity for pranking. A coat of white grease around the waistband or inside the crotch of a pair of powder pants would start to create a slippery situation after about 20 minutes of working after the break was over, especially in the summer. Grease inside of someone's work gloves was also a nice joke to watch unfold as a hand was squished into the goop. Powder clothes also came under attack in the form of knots. A tight knot in someone's pants leg tended to slow them down while they tried to get back into the swing of working on their standard for the day.

Another good target for grease was the occasional company truck driven to the line by one of the Foremen from the front office. Once, a fairly new Foreman drove a truck up the hill and then started to walk the line. While he was gone I decided to go grease the door handle so that as he opened the door, he'd get into the grease while I watched from a safe distance. However, as I greased the door handle, I decided that the gear shift, the emergency brake release handle, the radio knob, and a couple of other places needed a bit of lubrication.

Once finished and out of sight, I and a couple of cohorts watched the young Foreman as he eventually came back from his trip up the line and got ready to leave. I suppose being new to the treat he was about to receive, he hilariously proceeded to get into each grease trap and wipe it off, one at a time, until he hit about the 4th one. After getting a hand full of grease again he jumped back out of the truck and looked up and down the line. Then he started yelling that if he found out who greased his truck he was going to fire him. He ranted and raved and we were laughing at the frustration he must have felt. Still, it was done in fun and the next time we saw him on the line he laughed about it with us. Then one of the guys hollered at him from the Line Foreman's shack and told him someone wanted him on the phone. Of course, when he placed the receiver against his ear, he filled it with grease. He soon learned to be very wary around the line guys.

The Fight Club

For a bunch of grown men, and you had to be at least 21 to get a job there, we got a lot of school-yard types of punching and wrestling in during the course of a regular work day. There were about a dozen guys in our group who liked brawling. It was all done in fun, but sometimes it was not without a lot of pain and bruising.

The basic rule was no hitting on the face or head, or the groin. My arms and legs were almost always spotted with black and blue marks, as were the others in the group. Once in a while someone would get mad and then you'd hear the phrase come from someone else, you might be able to whip one, but you can't whip us all.

If someone did get over-excited and start to get more violent than the unofficial rules allowed, he would usually find himself tied to a porch post in front of one of the buildings until he cooled down. After someone got mad, which was rare, he was no longer welcome to participate unless he initiated the action, which usually resulted in a pretty good thumping just to see if he was actually back on board with the play.

Once, while I was battling with one of my good buddies at the Mix House, I took a big swing at his shoulder and he ducked, causing me to clock him right on the chin. I could tell by his red face that he was immediately more pissed than was good for me, so I grabbed him in a bear hug and started yelling for help from the Trucker coming up the line. After a few minutes and some cooling off time, all was well again, but you had to be careful. It was great fun and stress relief. Still, once in a while we would all call a truce and allow all the bruises heal and make it safe to walk between the buildings again for a few days.

The Skunk Event

As mentioned earlier, I was on the early shift and came in an hour ahead of the rest of the line. One winter morning as we walked up the line towards the Mix House we ran across a skunk. The skunk went into a culvert and we ended up killing it. Once it was dead, and smelling very bad, we put it in a bag and carried it down to the Case House. We hid it inside an empty box near the bottom of a stack of cardboard boxes the Case House put the sticks of dynamite into. Then we cranked the heater fan up to high and left.

Soon after the late shift came up the hill, we got word at the Mix House to slow down on mixing our day's requirements because the Case House was unable to start work. Eventually, the Day Shift Foreman from down front had to come up the hill and find the boxed-up skunk inside the Case House. Apparently, the Case House guys wouldn't even go inside the building. Then one of the union reps told the Foreman that he couldn't take the dead skunk off the line without getting a grievance filed because that was the job for the yard crew down front. Then everyone still wouldn't go back to work, or start actually, until the maintenance guy from the yard crew drove up to the Case House with an end-loader to carry the dead skunk back down the hill and bury it. The Case House refused to start work until the stink aired out from the Case House and the line couldn't start until over an hour later.

Not surprisingly, the Day Shift Foreman was highly irritated and angrier than anyone had ever seen him get before. He said he knew it was the early crew who had done this and that if he could prove it, he would fire us all. In hindsight, I can't blame him, but at the time we all thought it was hilarious and were thankful that being in the union made it really hard to fire us.

Fun and Games

Some days on the Powder Line were literally more fun and games than work. I've mentioned that being a union job, and due to the nature of the job we did, we could only mix the amount of explosive product that could be packed and finished in the same shift. We didn't have a night shift on the Powder Line side of the plant.

Working in the Mix House started with the Foreman telling us what types of mixes and how many he wanted to run for that specific day. We would start mixing by making sure we had enough to get started in each active Pack House once the workers showed up an hour after we did. Sometimes, especially on a Friday or even a Monday, some of the workers would simply not come to work. They enjoyed their weekends. The Powder Plant didn't pay as well as the coal mines in the area, but it was the next highest paying job around and some of the guys seemed to believe that three or four day work-weeks put enough money in their paychecks to get by.

So, with a high of maybe 45-50 mixes in a day, sometimes we would get cut to only 15 or so for the eight-hour shift. The way the system worked was that once you had produced your required standard and cleaned up your work area for the next day, you could go to the break house, which was also the Foreman's line shack, and as long as you stayed on the job site, you got paid. Sometimes I would be finished after only two or three hours. On days like this, the Mix House would usually help the Trucker and when we were all caught up, we'd do various things, depending on the weather and on what was going on.

We would play horse shoes and had a nice horse shoe pit set up with two lanes. We would sometimes play UNO, or Spades, or Yuker, or even Crazy-8. Some of the card games would last for hours.

The mules were kept corralled near the Powder Line and when off-duty, had a small field in the woods and a nice barn to stay in. They got fed very well with grains and hay. The local deer, somehow understanding that this was a no hunting zone, also stayed on the Powder Line. The mule feed was also fed to the deer and at times we would have 20 or more Whitetail deer eating mule feed off of boards stretched out along the ground.

Of course, with mule feed, you also had rats. The rats would build tunnels around the barn and sometimes we would carry buckets of boiling water from the steam lines running along the Powder Line and pour the water down the rat holes. This killed a lot of the rats, but often they would also run out of the holes and we could throw rocks at them or club them with sticks to help lower their numbers.

And then, there was also mule training. The Trucker, me and one of my friends John, were all quite comfortable around, and knowledgeable about horse and horse training. This played right into mule training, at least until we got a beautiful and big Missouri Mule up the line to break in. This was not our job, by any means, but the Foremen knew we had the time, the ability, and enjoyed working in this area as just something extra. As it turned out, this big mule happened to be the one that ended up causing my friend John to get injured and then the mule was sold off.

But, we had a lot of good times otherwise and it was hard work when we were working, but when we were finished, we really enjoyed ourselves.

The End of My Job

I got laid off and never rehired sometime in the summer of 1982. The Powder Line had shut down about six-months prior to my getting laid off and I had started working on the Gelatin Line. As a Utilityman, I worked a lot in the "Jelly" Mix House, which was a lot different than the Powder Line Mix House. Much more NG and larger mixes. I also worked quite a bit in various Pack Houses and these also were quite different. The mixes were not powder, but more like mud. You mixed the dope and NG in a large metal tub with mixers that would lift out when it was finished, then you would dump the tub into a large cart, all mechanically.

The Pack Houses were actually called Stuffer Houses. The gelatin was cut out of the carts with a small wooden shovel and then placed into something that looked like a sausage grinder. The cardboard tube would be placed on a nipple and the grinder would fill it up. As it filled, the next tube would be placed on the nipple, one after the other. These were the buildings that tended to blow up.

For whatever reason, production requirements slowed and the company laid enough people off that I was also let go. About a year later I decided to quit being a reservist in the Army and to go back to active duty. I enjoyed my time as a Powder Monkey and often think of some of the good times and good people working at Austin Powder provided.

During my time at Austin Powder, or most of it anyway, I was known to be a bit rowdy and somewhat of a trouble-maker at times. I was rough around the edges to say the least. I had left the Army early in 1979 as a Sergeant in the Artillery and got a little out of control with my sudden lack of responsibility I found at a job that required more brawn than brain, and where the only real responsibility I had was to do my narrowly focused job and do it safely.

With a little blame to the crowd I mixed with, but mostly on myself, I became somewhat of a bully and even more hard to get along with than I had been before getting hired by the Powder Company. My friends were close to me, as was my family, but everyone else became even more distant. I was not a generally nice person. However, that all changed.

In June of 1982, a few months prior to being laid off from the plant, I renewed my relationship with God and became very involved with a personal immersion into a self-managed Bible study and attempt at becoming a better Christian. By August 1983, I had gone back into the Army in a move that I felt was best for my family and our future. As it turned out, you can take the boy out of the dynamite, but you can't take the dynamite out of the boy. My future ended up with me as a Demolitions Sergeant as a Green Beret when I volunteered for Special Forces training and really changed my life once again while creating opportunities for many more extraordinary tales.

BICEPS, a Short Journey in Small Business, Albany, OH, 1983

I partnered with Rob Bricker, a good friend and the pastor of the church we were attending, and started an unlicensed electric and plumbing business. With a couple of jobs under our belts, we became BICEPS, which stood for Brothers In Christ Electrical and Plumbing Service. We did primarily re-wiring, which is replacing and improving old electrical wiring, and a few small plumbing jobs. We worked in unincorporated areas of the county to avoid the need for business licenses and insurance.

This lasted about a year, but we could never get enough work to stay busy or financially stable. I enjoyed the work, but besides there being no business insurance, there was also no family insurance, medical or otherwise, and again, the work wasn't really steady.

It was about this time, the summer of 1983, that Linda and I discussed going back into the Army. By now I was missing a lot of the comradery and professionalism that I had found with the military, especially when it was compared to the civilian work I had been doing. And besides medical insurance, I knew the Army presented a steady paycheck. I had a lot to consider as the dad of a quickly growing young girl and a wife who I was responsible to, for providing them a good home.

Likewise, there was the patriotic aspect, which also weighed heavy on my decision. Soon, we agreed that the Army would be the best choice for our next chapter in life.

My Motorcycle Days, 1979 – 1983

It was the result of several things. Gasoline was expensive. A new car was even more expensive and I'd been having a hard time keeping my AMX running due to the transmission that kept going bad. Also, Linda needed a vehicle since we both were working and our jobs were in different locations. Economics played a part, that's for sure.

Then there were my work buddies who liked to ride dirt bikes. I wanted to ride with them. It sounded like a lot of fun and I'd always thought it would be something I'd enjoy. I'd liked riding my mini-bike I had for a summer (when it worked). Also, one of my good friends had used to race motocross and he had a great time doing it. The fun aspect was definitely a major factor in my decision to get into motorcycling.

The bug had bit me and I soon went to the local bike shop and came home with a brand new, 1980 model of a Yamaha Enduro 175. A nice 2-stroke, street and trail motorcycle. I rode this bike everywhere and had a great time. Then, winter arrived.

I remember riding this bike to work when there were several inches of snow on the highway. Lucky for me, I was on the early shift and usually, there were few tracks in the snow on the roads I traveled to get to work at that time of the morning. I could usually get about half-way there before having to stop and thaw my hands out on the exhaust manifold. Some days it was a real adventure, but I always made it there and back.

Not long afterwards, sometime the next spring, I broke down and got a road bike to ride to work. I had an old yellow Volkswagen Bug that I traded for a 1975 Triumph Bonneville 750. This bike had been partially chopped and had a slightly extended front fork and a modified exhaust that you could hear coming down the road for a couple of miles.

I continued to enjoy the motorcycle game and ended up going from a 175 Yamaha for playing in the dirt, to a 250 Suzuki, and then to a 400 Husqvarna. I remember the dealer telling me to be careful with the Husky because according to him, it would roll my socks up and down. It was a powerful dirt bike.

It wasn't too long and I was out riding with a buddy of mine when my socks got to rolling. We had been in the woods on some trails and in some mud. We came to a place where we were going to follow a gravel road for a bit as we headed for the trails again.

As I started around a slight curve to the right, I hit some deeper gravel and between my speed and desire to stay in the lead, my bike got a mind of its own and decided to skip across the road and into the ditch on my left.

I hit the ditch knowing what was coming as I tried to keep from killing myself.

The first thing that happened was my quick trip over the handle bar as the front of the bike became a plow and the rear wheel raised up into the air. I probably should not have hung on so tight, but there wasn't much time to think about it. I was cleaning mud out of the ditch with my helmet and facemask before I could say ouch. Quickly, as the bike started slowing down, I thought to myself, that wasn't so bad. And then, the rear tire touched down.

When the rear tire came back down into the ditch, I was launched yet again, back along the ditch for what seemed a long way before it finally stalled out and everything stopped moving. I was alive, but I wasn't sure how well I'd held up during the trip. I began to mentally take inventory of my body parts.

In what seemed like a minute or two, but had to be only a few seconds, my buddy came riding up and jumped off his bike to check on me.

He seemed worried enough and asked me if I was OK. I had figured out by then that I was not really hurt, other than my pride, and told him I was good as I started getting untangled from the bike, the mud, and whatever else had become attached to me as I surfed down the ditch.

My buddy then told me, "well, if you're OK, then that's the funniest thing I have ever seen!" and started laughing like he meant it. I had to admit, it was pretty funny. Especially since I didn't get more than a couple of scrapes and a bruise or two. The bike was fine also, even though it wouldn't start for a while. Probably being upside down for a few moments didn't help matters.

That was the worst crash I ever had on a bike, although I had several upsets on the trails during our various rides.

I came close to biffing it on the roads a couple of times, but managed to keep it upright in in the highway. Once I had a wasp fly up a sleeve and sting me about four times before I could get to the side of the road and pull over. That was no fun at all and a thousand wonders I didn't crash. Another time, the Triumph popped an unexpected wheelie as I passed a car leaving work a bit too slow for my liking. That was not only unexpected, it was a rush (and scary) as I finally got the front wheel back down and the bike back into my own lane. It probably looked cool, but I didn't try it again.

During the motorcycle period, even Linda got her license. She would ride the Yamaha while I'd ride the Triumph and we would cruise the back roads and less-travelled highways. We really enjoyed the bikes, but once I went back into the Army, I never got another bike. Now I watch my grandkids ride once in a while. I keep thinking it would be fun to get another bike, but that's about as far as it goes these days.

My biking is now limited to my mountain bike and still, even peddle-power in the gravels causes me to remember my lesson learned on the crazy Husky so long ago. I don't want to clean out any more ditches.

Daddy's Shadow

Lori was still a tiny baby when I left the Army and Linda and I moved back to Albany, Ohio. We spent a month or so living in Linda's childhood home and sleeping in her old bedroom. We enjoyed it well enough, but Lori didn't go to sleep easily at night and liked to tell the world that she was still awake in long and loud sessions of crying. I soon found something that helped.

Mommy was working, so she needed some rest at night. Papaw and Mamaw were, well, great-grandparents and also needed some rest at night. I'm not sure if they got much rest until Lori went to sleep though. Through desperation as much as parenting skills, I soon found out that if I held Lori and walked in circles, rocking her in my arms, and singing louder than she was crying, she'd soon calm down and go to sleep. "By oh by oh baby" sang over and over seemed to be her favorite tune. She would either relax or give up and drift off to sleep with the song, if you could call it that. At least she'd feign sleep until I tried to lay her down in her crib and then, we'd start all over again. That didn't last long though, and soon enough, we were all able to sleep at night without the barbaric serenade, her a-crying and me, a-screeching.

Within a few months, I had repaired and upgraded a small three-room building that used to be a storage shed into our newest apartment. It was part of Papaw's garage, very much a shed, and very rough looking. We moved to this small space across the driveway and continued our transition from the military life to a new and quite different civilian lifestyle. Soon I was employed and about that time we bought a nice, albeit used, mobile home. A 14 x 60 foot mansion on wheels. We had it pulled to a nice five-acre section of the farm that Papaw had given to me and Linda. This became home for our small family for the next three and a half years and became Lori's new playground.

My daughter Lori was my pride and joy. She was a smart kid, but what really set her apart was her seriousness. People would always comment on how well-behaved and older-than-her-age she acted. As I worked around our property, most of the time Lori was right there with me. Her hair was almost white from being in the sun so much and you could see her head bouncing just above the hay in the field as she followed me from one spot to another when we were working on the fence, going out to catch our horses, or hunting grasshoppers to go fishing.

At a very early age, I taught Lori how to ride a horse. Her horse was named Star, a young mare I had broken not only to saddle for riding, but also to harness to help with the work around the place. We had gotten Lori a pony for a while, but the little pony was meaner and more trouble than either of the horses were. Soon, Lori was riding Star in small horse shows in the area. We'd also go on trail rides and Lori would double with me as we traversed creeks and hillsides.

Besides horses, I also used to ride Lori on my motorcycles with me. She had a sparkly blue helmet and would ride in front of me, holding to the tank while I held her legs against the bike with my own. I was much more carefree, and either braver or more foolish as a young dad than I became when I reached the grandfather age. But, we never had a problem and Lori loved riding the motorcycles with her daddy and visa-versa.

As Lori was turning five and getting ready to start Kindergarten, I went back into the Army. One of my saddest memories was watching Lori wave goodbye as I left Athens on a bus headed for Ft Dix to be processed back into the military. I can still hear her tell me "goodbye daddy" as I gave her a kiss and promised I'd see her again very soon.

It wasn't too long and she and her mom were with me again in Germany. It was good to have our small family together once more and within a year, we three became four as Joshua was born. Before long, I had two shadows. However, living on a military base where there were many other kids, my shadows were kept quite busy with so many friends and more fun things to do.

But for a few years, on a small hilltop in Albany, if you saw me walking somewhere in the field or around the barn, there was most-likely a little blond-headed shadow following close behind. Those will always be some of my favorite days.

You're in the Army Now, Again, 1st BN, 36th Field Artillery, Germany

Once Linda and I decided that the Army life was the life for us, again, it didn't take long to happen. I went back to a recruiter and after a four-year break, within two-weeks I was back in the Army, back in the Field Artillery, back in my old job as an FDC Specialist, and even back in Germany where my Army career had originally started in 1975. The only change was that now, instead of being a Sergeant, like I'd been 4-years earlier, the Army made me come back as a PV2 and I was wearing that little mosquito wing again.

But, I held my head high and I got back into the swing of things really quickly. With my former experience, as a PV2 in my new unit, the FDC Chief quickly gave me the senior position for computing the firing data of the 8-inch self-propelled howitzer unit. This was the same type of unit I had been assigned to the first time I was in Germany, eight years earlier.

Within two years, I had been promoted not only to my original rank, but to the next grade, Staff Sergeant, E6. But when I first arrived, within a month, Linda and Lori had come to Germany and we lived in the city of Augsburg, about a mile or so from Reece Kaserne, the US Military post to which I was assigned.

During our time in Augsburg, we decided that it was time to have a second child and in September, 1984, our son Joshua was born. After two-years in Augsburg I was reassigned to Ft Bragg, NC with a stop at the Army's Airborne School enroute. Jump School was an awesome experience, but one I would only want to experience once.

After I graduated and signed into the 18th Airborne Corps Artillery at Ft Bragg, the Army was starting to really become exciting and fun again. Being a soldier was nice, but being an Airborne Paratrooper was on a totally different level. Things really started to change for me and my family once Ft Bragg, North Carolina became home.

Family Life in Augsburg, Germany

But, before I got to Ft Bragg, I spent two years in Augsburg. My second tour in Germany was better than the first one had been. Regretfully, I started my 2nd tour wearing the same rank as I did the first tour because of my break in service. As luck would have it, or perhaps karma, one of the first people I met in Germany this time was an old friend.

I say regretfully because when we were friends in Baumholder, the first time around, we were both Sergeants. E5, Buck Sergeants and we were in the same unit and were good buddies. He and his wife lived in downtown Baumholder and after Linda returned to the States a little early, I used to go to his house and visit him and his wife. It was a nice break from the barracks.

This time around, after being in Augsburg a while, Linda and I took Lori and went to the "big" PX located in Stuttgart on Panzer Kaserne. While there I met my old friend Phil and his wife in the snack bar. It shouldn't have made any difference, but now he was a Sergeant First Class (SFC) and of course, I was a PV2 again. I was uncomfortable because of the circumstances, even though I shouldn't have been. I suppose it was a reminder of what I had not done with regard to the Army, because of my break in service. I forgot the good things that my break of four years had allowed me to do.

I never did get back together with Phil after that and I only bumped into one other guy in Augsburg that I knew from my previous enlistment. When I was reassigned after my time in Augsburg, I went Airborne and then to Ft Bragg, eventually going to Special Forces and I left the Artillery circles all together.

We lived in an apartment building in Augsburg that was in town and not part of the military housing area. It was within walking distance of my unit on Reese Kaserne and I did walk once in a while, but usually drove my car. I could also take a bus if I wanted to go that route and it would stop very near the main gate.

Linda and I really enjoyed Augsburg. Granted, I was gone a lot, but when I was home, we had a good time. We would ride the trolley car from just a block or so from our apartment into the main downtown area and there were many historic sites to visit, large department stores, and some very good restaurants.

One of our favorite ones was much closer to our apartment and we went there about every month or so. It was a little pizzeria and they made personal sized pizzas. We would get a pizza the same size as the plate it was served on and a nice salad. Lori liked the pizzas also and that made it even better.

Speaking of Lori, the military school had a bus that would come by our apartment and pick her up and bring her home every day. She attended the American elementary school on one of the other Kasernes in Augsburg. Life was good.

Bike Riding

One of our favorite pastimes in Augsburg was riding our bicycles. Linda and I had bought a couple of standard German bikes and took advantage of all the easy bike trails, paths, and bicycle lanes of the city streets. I put a child carrier seat on my bike and Lori would ride with us whenever we went out.

Talking to Lori, the bike rides are some of her favorite memories of our time in Augsburg. She would set behind me in her bike-version of a safety car seat. She would always give the turn signal with her hand whenever we were changing lanes or directions.

We would ride for miles in and around the city and pretty much never be out in the street with traffic. The bike paths went through the parks, along a small river that wound through the city, and you could go into the countryside or the main downtown areas.

After Linda became pregnant with Joshua, she continued to ride all through her pregnancy. I believe the last time she rode while pregnant was the time at about nine-months along, she lost her balance and tipped over at one point where we had been forced to slow way down along a section of a street running under a bridge. She didn't hurt herself, but I don't believe she rode again until after Josh was born.

Josh is Born, Lori Becomes a Big Sister

Josh was born in the Augsburg Army Hospital. I was home when it came time, and took Linda to the hospital. I remember how proud Lori was to have a little brother and she quickly learned how to help take care of him and soon became a little mother to him. Lori being six-years older than Josh was old enough to be a big help. After Josh was about five months old, we decided that Linda and the kids would go back to the States and visit with family for a couple of months before I departed Germany and headed for Airborne School.

A couple of our German friends, a girl who was dating one of the former soldiers in my unit who had got out of the Army and stayed in Augsburg, took us all to the airport in Frankfort for the trip back home. We had a good time at the airport and stayed until Linda and the kids flew out. Then we all went back to Augsburg and I started moving out of the apartment once the packers got everything boxed up and shipped stateside.

I went to the barracks and by then was on the promotion list for Staff Sergeant, so I got my own private room. I was in the field much more than I was in the barracks, so it made the time go by quickly until I was ready to leave Germany once again.

I was soon on my way to Jump School at Ft Benning, GA and life would never be the same again.

The Karate Kids

Life's occasional scrapes with a bully (although some thought the bully was me), and some Special Forces combatives, or what was also known as "Hand to Gland" training, provided me with a working knowledge of how to survive a fight without the advantage of having a firearm. However, I wanted something more structured in the realm of self-protection for my children and martial arts seemed to make a lot of sense.

About the time that Lori turned 12 and Josh was 6, I got them started in Karate. They both enjoyed the sport, as well as the discipline that came along with learning the techniques. They also enjoyed belonging to a dojo with others in their group. Lori and Josh did well and Lori earned a Red Belt while Josh had earned his Blue Belt in Karate.

Besides earning rank and doing well at the tournaments in their displays of the various routines, both of them were exceptional in the fighting competitions as well. I think they liked the idea of a contact-sport along with the challenge of physical competition. I have to admit that I enjoyed seeing them knock their opponent across the mat also.

About this time, we decided to switch to Taekwondo. The Taekwondo school's owner and primary instructor was a Korean-American woman who had competed in the sport at the Olympics and was a very respected person in the sport. Once again, both of the kids excelled in learning the techniques, learning to use the various weapons associated with the sport, and in the fighting competitions at the many tournaments we attended. It was not long before both of the kids had earned their Black Belt status, in fact Lori even earned her 2nd level Black Belt in this new discipline of martial arts. While both seemed to have a natural talent, Lori excelled in the fighting areas as she grew older and stronger and very quickly became a top contender in her age group. The last year she actively trained, she won her class in the NC State Championship tournament and was selected to attend pre-Olympic tryouts after her next win at the regionals.

However, my change in assignments from Ft Bragg to Yuma Proving Ground, AZ curtailed her Taekwondo career and once we moved, she no longer focused on Taekwondo. Soccer soon became her sport and as a goalie, she continued to enjoy the roughness of the sporting combat. But while she was in Taekwondo and actively participating, she struck fear into many of her opponents when they met her on the mat and squared off before the fight.

Without exception, both Linda and I attended not only every tournament, but almost every practice session at the school. We could be heard cheering them both on in practice or contests, as they participated in all the events. Our home soon filled to the point of overflowing with trophies and medals. We were likely considered obnoxious as we would loudly cheer and yell, even if we were expected to remain quiet. These are some of the most memorable times I enjoyed while I was home and able to share these hours with my kids, even if I was only there as an observer and not an active participant.

I also knew that besides having fun, getting some great exercise, and learning new skills, they were being prepared for the possibility of having to defend themselves or someone they loved in a bad situation. The confidence it provided by simply knowing that a fight would not break them, was instrumental (I believe) in setting the stage for the high level of success they both experienced as they grew up into adults and parents, as well as contributors to the general well-being of not just our nation, but humanity itself. I am proud to see my greatest contribution to the universe has been my children and what they have achieved and produced, to include their own children and their successes.

Learning to Fly

Josh was still in a diaper the first time I strapped him into a parachute harness and let him leap off a platform 15 feet above the ground to slide along a cable towards the cushioned wooden pole at the end of a zip line. We didn't even know it as a zip line back then. For us, it was a kiddie version of the 34-foot Jump Tower that was a part of week two, or Tower Week, during the Army's Basic Airborne Course at Ft Benning, GA.

Both Josh and Lori had many jumps from the mock towers before they were teenagers. The great thing about going to the fair at Ft Bragg, was the Jump Tower and the Rappel Wall for kids. When you grew up at Ft Bragg, NC, you grew up thinking everybody's dad jumped out of airplanes and that jumping off a tower above the ground, or walking down a vertical wall dangling from a rope was a normal thing to do.

An additional perk for my kids came after I was assigned to the Military Free Fall School (MFFS) as an instructor. Ft Bragg had a huge wind tunnel in which the students got to train in simulated free fall. A giant fan blowing the wind up from beneath the flooring propels the student into the air where the art of flying is taught, and controlled movement in the air is experienced.

I brought both of my kids to the tunnel and allowed them to learn to fly on a cone of wind that simulated free falling from the sky. Lori was a little easier, as she was no smaller than some of the light-weights that attended the MFFS as a student. She learned how to arch and use her hands to turn left or right and was able to adjust her level by arching or relaxing with little effort. Josh, on the other hand, was still much lighter than our normal light-weights. As Josh entered the airstream, if I had not had a good tight grip on his jumpsuit, he would have shot straight to the top net high above the level from which we entered the airflow. In his yellow jumpsuit, he looked all the same as a Cheeto that had been tossed out a car window going down the freeway.

Still, with a little practice, he was soon turning and elevating or lowering himself with little or no direct contact with dear old dad. They both managed to get about 100 sky dive jumps worth of free fall time while we played in the wind tunnel. For them it was play. For me, it was actual experience working with a student who was learning to fly for the first time.

Then, once they both were older and after I had retired from the Army, they went one step further. Both went to a civilian sky dive business in Colorado and completed a tandem-jump with a former MFFS instructor and Teammate of mine. He had also retired from the Army by then and was working as a Tandem Master at the sky diving school and airfield.

Of course, I was a proud dad once I found out and watched the DVD of their jumps. Lori had a rough landing, but they both survived and lived to talk about it. So, I had the privilege of watching my kids go from crawl, to walk, to run, to fly. Not too many parents watch that process unfold and I'm one of the crazy ones who probably pushed the inspiration out on purpose.

Fast forward 30 years. On a mountainside in the Rockies, two of my grandkids, eight-year-old Kody and six-year-old Kerrington, climbed about 20 feet up a crude ladder fastened to a huge pine tree, strapped on a Swiss Seat and helmet, and jumped off a platform to zoom down a long zipline headed toward the valley floor. Mom-Lori and dad-Stew, along with Linda and I, looked on as they both shouted out AIRBORNE or something loud, smiling from ear-to-ear as they defied death and gravity. Genetics or environment? They say crazy doesn't fall far from the tree. Airborne!

Geronimo – Airborne Paratrooper

Geronimo! That was a Paratrooper's war-cry as he left an aircraft before all the politically correct restrictions started impacting the military. Still, being Airborne qualified and on Jump Status in the military was an identifiable bump up in status amongst all other military skills.

I had to write a chapter on my Airborne experiences. With 663 parachute jumps, and at least a little information about each one written down in one of my four log books somewhere, I knew I could find several of the more exciting ones about which to write a short paragraph or two. Not a lot of people jump out of airplanes, and even fewer jump out of jets, in the dark, wearing 250 pounds of light-weight equipment, from up to 30,000 feet above the ground, sometimes in foreign countries, with a group of equally crazy companions. It should be an interesting chapter, especially for those who never have jumped even once. Of course, being indoctrinated with the elements of superiority that Airborne Paratroopers exude, especially among non-Airborne (leg) personnel, both Lori and Josh had to try it and each has made at least one exit from an aircraft while in flight via a tandem jump. As far as I know, at the date of this writing, neither have ever felt compelled to do it again, but that's OK. They have jumped at least once. Doing so, they have actually done something I never did, and that was make a purely civilian parachute jump. I have never made a non-military related jump. A few that were military AND a skydive, but still had a military duty purpose. And even with all those military jumps and experience, I never felt like I wanted to jump after I retired.

No brag, just fact. I was a very highly qualified and experienced Paratrooper. My military qualifications included the Novice, Senior, and Master Parachutist levels for static-line parachuting. I also held the Novice and Master Military Free Fall ratings. Additionally, I was rated as a Military Free Fall Instructor (MFFI-336) and a Military Tandem Master (MT-17).

I had a few foreign parachute wings also. I was awarded the Master Parachutist wings from Jordan. Also, the Army Military Free Fall wings from Zaire. From Zimbabwe I received the Army Parachutist wings, the Special Air Service wings, and the Air Force Parachute Instructor wings.

Of my total parachute jumps, 608 were MFF jumps with a total freefall time, that is falling through the air before I opened my parachute, of 9 hours, 54 minutes, and 27 seconds. Most of the jumps were as an Instructor at the Military Free Fall School, advanced skills battalion with the Special Warfare Center and School, but I also had about 260 MFF jumps to include both HALO (High Altitude Low Opening) and HAHO (High Altitude High Opening) also known as Stand-off jumps with my assorted A-Detachments or MFF SFODAs, 594, 334, and 384.

As a static-line and MFF Jumpmaster, I could plan and execute all types of military personnel airdrop operations, to include bundled equipment for parachute insertions.

I was also rated through the civilian side of the parachuting community via the United States Parachute Association, where I held a "D" license, the highest rated skydive level for civilian jumpers. This allowed me to do stadium jumps and other risky or small drop-zone jumps.

The following are some examples of the varied and sometimes difficult Airborne operations I've completed.

My First Parachute Jump

I attended the Army's Basic Airborne Course at Ft Benning, Georgia during October, 1985. I was a Sergeant (E5) at the time and as an NCO or November as the Cadre referred to us, I held a special place in-between all the new privates and the new officers, mostly 2LTs, who were also in my class. Besides all the training, we few NCOs were required to manage the unit in all but the actual Parachute School training functions. This added burden offered me ample opportunity to spend extra time in what was known as the "gig pit" doing exercises and enjoying some Airborne Instructor, or Black Hat personal attention. This was the type of school that inspired everyone to give 100% and then some.

The school was basically broken down into three weeks. Ground Week, Tower Week, and Jump Week. Here we all learned the fine art of being able to safely land on the ground after falling from various heights and conduct what was called a PLF, or Parachute Landing Fall. Instead of trying to catch all of the impact on our feet and legs, we learned to spread the impact across our bodies by landing with a side roll that allowed our feet, calves, thighs, buttocks, and back muscles to absorb the shock and prevent injury. We also learned how to wear the parachute assembly, how to exit the various aircraft doors, and how to collect our equipment after landing on the drop zone. While the detailed instruction was very capable of getting the new paratrooper ready for Jump Week, nothing but the actual 1st jump could define what it meant to exit an aircraft while in flight.

If forced to only use one word to describe the first jump experience, I would have to say violent.

You grow up watching Bugs Bunny and many others make parachute jumps in the cartoons. Maybe you watched them jump in some D-Day movies, or skydiving shows on TV. Even watching the actual jumps take place from the ground gives you a faulty sense of what the real experience of moving from the inside of the aircraft to the outside is like. It's much like stepping into the center of a tornado.

My first jump was made over Fryer Drop Zone or DZ on Ft Benning, GA. It was made from 1250 feet above the ground and from a C-141 jet powered troop transport. I was in the middle of the stick, which is what the group of lined up paratroopers waiting to get to the exit is called. Our static lines were attached to the cable running just above our heads and this was attached to the top of the parachute we were wearing on our backs. As we jumped out, the static line would drag our parachute out and allow it to fill with air as we fell away from the aircraft, breaking loose as our risers were completely extended and we would then be floating towards the ground.

As I've mentioned, the biggest surprise I had was the level of noise and physical impact from the wind that I felt as I left the door and leaped into the sky. It passed quickly though, and within three to four seconds, all was calm and quiet again as I drifted downward. About that time, I suddenly remembered to check my canopy and look around for other jumpers, the ground, and where I was headed. The contrast between chaos and peacefulness can't be overrated. Still, that's only if all goes well and there is no emergency to deal with. I enjoyed my first jump all the way to the ground. That's when things changed.

It was a bright, sunny day for my 1st jump. The wind was blowing a little, but not too much. About 6-8 knots. My exit had been good, if loud, and my opening and trip to the ground had been about as perfect as it could be. Again, other than the surprise at exit, the ride was about like the one I had gotten from the top of the 250-foot tower the week prior. But, as I drifted down and looked toward my landing spot, I saw that it was going to be at the bottom of a slight dip in the open field that looked like a dry creek bed. It seemed to me that at the last second or two, my rate of descent increased dramatically and smashed me into the dirt like a ton of bricks. I did my PLF but immediately thought that if I hit the ground this hard every time I jumped, I was not going to make it through the week. It seemed that every bone in my body got jarred and I ached all over more than anywhere else. Suddenly, I was not looking forward to the next jump, which was to come early the next morning and would include strapping a rucksack and a rifle to my body as an extra bonus.

The next morning saw us loaded into a C-130 Hercules and headed out to another trip across Fryer DZ. Once more, I was somewhere near the middle of the stick and this time as I exited the aircraft, the whirlwind was almost ignored as I felt the opening shock of my canopy and concentrated on checking everything, lowering my rucksack on its lowering line, and landing. The landing was a huge relief. Even with the rucksack, I landed much more softly than I did on the first jump. In fact, the remaining jumps at the school were all followed by much easier landings than the very first one. I was graduated and on my way home after jump number five. While still on the DZ I got my Airborne Wings pinned to my chest and a big boost of self-assurance pinned to my ego. I was soon headed to Ft Bragg, NC and on to having a great Airborne career.

Jump 7, Airborne Infantry

After Airborne School, it didn't take long before I discovered more of what being Airborne actually entailed after leaving the sterile and quite safe environment of a school, very closely and professionally controlled by Black Hats who were always totally in control of the process from start to finish. As they say, welcome to the real world.

The 82nd Airborne Infantry Division, 504th Infantry Regiment was a rowdy bunch of paratroopers with a long and distinguished linage dating back to WWII. Nothing much had changed and on my 7th jump I was going to be sharing the air with over 600 of these wild and crazy paratroopers, at night, strung out behind a long line of C-141 aircraft flying low and slow over the Normandy DZ on Ft Bragg, NC. There are very few experiences equal to this.

This jump, called a Mass Tac for mass tactical jump, started long before you climbed aboard the aircraft at Green Ramp on Pope AFB just outside Ft Bragg. While Jump School had the students jumping nice and neatly packed square rucksacks, each weighing 35 pounds, and a weapons case with a wooden 2x4 the length of an M16 inside, getting ready for a real-world jump in the 82nd Airplane Gang was a horse of a different color. Rucksacks were loaded for bear and many reached near the 100-pound mark and they came in all sorts of oval or oblong shapes. Weapons cases may contain M16 rifles, machine guns, or sniper rifles. Some weapon systems didn't even fit inside a case and were padded and strapped to the jumper, such as the AT4 anti-tank rocket, mortar tubes, or other various weapons of war. Of course, ammunition was included as well, and water, chow, or other essential items like radios, night-vision devices, and batteries.

Knives were also evident in many shapes and sizes. Strapped to the lower leg, just in case you got towed behind an aircraft and had to cut the static line to get free. And no one paints their face like an Airborne Infantryman on a tactical infiltration. Camouflage paint was covering every bit of exposed skin on the head, ears, face, hands, and neck.

Before the doors ever open, the fun gets started. On tactical jumps the planes would always spend at least an hour flying low-level routes to the DZ. This was the USAF getting their credit for their combat training flights. Low-level means low-level in this instance. Flying just above the tree tops and following the terrain of the ground, the G-force exerted on the paratrooper setting in the back as the plane climbs over hills and banks into tight curves, followed by the momentary weightlessness as the plane dips quickly toward the ground again has caused many a strong stomached paratrooper to fill a barf bag or two.

Finally, we got the 10-minute jump command and started standing up, hooking up, and getting ready to exit. At ten minutes out, we were still flying low-level and the rucksack, web-gear, and parachute felt like it weighed several hundred pounds at times as we stood there waiting for the relief of getting our knees in the breeze. Then the doors were opened and it got really real. As we got close to the DZ we got the one-minute warning and soon after, the pilot flew the aircraft higher than the current tree-top level to 800 feet above the ground. This was only about half the distance above the DZ from which I had ever jumped at this point. From my position in the stick, I could look out a window and see lights on the ground, headlights of cars and eventually some lights at the edge of the DZ, but I couldn't really judge how close we were because it was so dark. At one-minute out I witnessed something I had never seen in Jump School. Suddenly everyone started jerking on the static lines and the anchor line cable they were attached to started bouncing up and down and the jumpers were yelling and hollering. It may have been bravado, but it sure pumped everyone up and it got my attention with whatever extra there was to give. It was truly like being inside a can of whoop-ass.

At the 30-seconds command, the number one jumper was told to stand in the door by the Jumpmaster. When that happened, the stick tightened up and the main guy responsible for that was usually a well-seasoned NCO near the front of the aircraft, who would be the last jumper in the stick. He was known as the stick-pusher. There was no lollygagging between jumpers and the time it took to get everyone out in one pass was determined by how long the DZ happened to be. On this jump the bodies were flying out the door until they all were gone, me included. If someone in the middle had picked up his feet, he would have still made it out the door we were so tight.

So, as it is said, when the green light popped, the BS stopped and the Jumpmaster slapped the first jumper with a loudly yelled GO! One by one, from both the right and the left doors, the paratroopers jumped, stumbled, slipped, or were pushed out the door and into the dark sky.

As I went out the door I could see the jumper in front and watched the parachute come off his back and open in spite of how dark it was. Suddenly there were parachutes all around me with at least two so close I thought we would touch, and in fact my canopy did bump into another canopy as we started to spread out and drift towards the ground. The ground looked close, and I could make out the difference between the DZ, the trees along the edges, and the horizon. By the time I checked my equipment and looked down again, it was already time to lower my ruck and get ready to do a PLF. Suddenly, in what seemed only a few seconds, I was on the ground along with more paratroopers landing, gathering their chutes, being dragged, or heading for the assembly point than I'd ever seen in one place before. That's about the time it hit me, I didn't just like this, I loved it. What a rush that had been. As I got my gear gathered and was heading towards the assembly point, the next sortie of aircraft was roaring past just over our heads. As they started to exit I heard a thump as someone's helmet must have been ripped off by the deploying risers during a bad exit or poor body position as he left the aircraft. Then a louder crash as someone's rucksack fell to the DZ instead of remaining attached to the jumper when he lowered it. I hurried along not wanting to risk getting hurt by something falling from the sky after I had made it through the jump.

I had more Mass Tac jumps, but none were as exciting as the very first one. I really felt like a paratrooper after that and I knew then that I wanted to be a Jumpmaster as soon as I was able to try. And a stick pusher. I wanted to do that also. I wanted it all and I wanted it now.

Jump 20, And Now, I'm a Jumpmaster

I admired the Jumpmasters of the unit. The star and wreath on top of their jump wings was a mark of honor in my eyes. I knew some of them weren't any good at their normal job, but they were good Jumpmasters and that seemed to be enough. Most of the Jumpmasters were true wing bearers, not just wing wearers as so many 5-jump chumps were all across the military.

I enjoyed responsibility and jumping was no different. When given a choice, I always preferred to lead rather than follow. That was both a strength and a shortcoming. I always felt that I'd be better at whatever it was that I saw one of my leaders or commanders doing and because of that, I sometimes got myself in a bind. Still, it was more a blessing than a curse, especially in the military.

I let my ambition to become a Jumpmaster be known to my leaders as soon as I was assigned to my unit and I started assisting when and where I could as we went through pre-jump planning and preparations. I was given a slot in the 82nd Airborne Division Jumpmaster School as soon as I had made the minimum number of jumps. I made my first Jumpmaster rated spot and jump on jump number 18 on the 4th of August, 1986. I'd only been airborne qualified for 10 months. After graduating from Jumpmaster School, I went back to the unit and started supervising some of the soldiers who had been jumping for years and had many more jumps than I did. I remember one in particular who thought he would get under my skin by calling me a Cherry Jumpmaster because he had over 50 jumps. I just laughed and reminded him that he was still not a jumpmaster, so he needed to do what I told him to do, cherry or not. I quickly earned the respect that good jumpmasters got from those who enjoyed jumping, but didn't want the headache, that is, the extra work and responsibility of being a jumpmaster.

My first jump out of the Jumpmaster School was made as the assistant jumpmaster for a parachute operation that was going to open an airshow at Dobbins AFB, GA. We flew down to Dobbins from Ft Bragg and made a Mass Tactical jump onto the airfield from a C-130 from 800 feet. I was on the right side of the aircraft making my safety checks outside the door and remember that my 30-second checkpoint was I-75, just about ten miles north of downtown Atlanta. I was the last one out the right side and landed between the main runway and the taxiway near the west end of the airfield.

We quickly assembled and turned our equipment in, donned our Airborne Paratrooper denoting Maroon Berets and then mixed with the crowd who was there to watch the show. I remember I even signed a few autographs that day and even though it was before the age of everyone taking pictures with their phone, there were still a few requests to be in photos for some people who brought cameras with film.

I had made my debut at being a jumpmaster and done so as part of the jumpmaster team for a very high-profile operation. The cherry jumpmaster stigma was gone and my status as one of the go-to jump operations team members was solid. Hanging out the door of the C-130 while flying above the interstate and lining up on the DZ that was surrounded by roads, buildings, and even aircraft on the ground was quite exhilarating. And to think, I was even getting paid to jump.

Jump 30, Robin Sage and Broken Leg

Of all the times to get hurt, jump number 30 as a student nearing the end of the Special Forces Qualification Course was really bad timing. This was a jump into Rudie DZ in the scenario-based country of Pineland. Real-world, it was a small field in the Uwharrie National Forest area in NC. It was a night jump with huge tactical loads in our rucksacks as we were going into an Unconventional Warfare mission for Robin Sage, the final exercise needed to graduate from the course. We jumped from the tailgate of a C-130 with a pass of five jumpers. The first half of my team had jumped during the prior pass. I was the stick pusher, although I didn't really have to push anything, as we were all anxious to get out and get started.

I remember as I left the ramp I noticed a flash of lightening way off in the distance and a lot of big clouds made visible for a second as the bolt lit up the very dark night sky. I thought that this was a bad sign. As it turned out, I was the last jumper to exit the aircraft that night and the other teams on the bird returned to the airfield and were trucked into their DZ assembly points because Airborne Operations got canceled because of bad weather just about the time I got under canopy.

I could tell the wind was blowing as I neared the ground and I saw that I was traveling backwards at a pretty high rate of speed. Long story short, when I landed I hit hard and apparently also landed on my rucksack as it was bouncing along behind my wind-blown chute. I felt a terrible pain in my leg and discovered soon enough that I had broken my fibula and dislocated my ankle when I hit the ground. I was not going to finish Robin Sage during this cycle. Still, even with the broken leg that twisted very abnormally every few steps, I gathered my gear and made it to the assembly point about a half-mile from where I had landed. When the medic looked at my ankle, the evaluator called in the ambulance for the long ride back to Womack Army Hospital on Ft Bragg.

I had surgery to fix my injury and two months and two days later, I made jump number 31 into the same DZ, with the same amount of equipment, but didn't break anything when I landed. Actually, I broke out a big smile when I landed and hadn't hurt myself any worse than I was hurting from simply still being on the mend. I kept the Motrin makers in business over the next month and graduated a little later than I had planned, but still sooner than many expected after the injury I occurred during my last jump.

Jump 41, Military Free Fall School

I really enjoyed jumping. I decided that if I got the chance, I wanted to go to Military Free Fall School (MFFS). We called it Halo School back then, but it was technically MFFS. I had been in 5th Special Forces Group for several months and got the opportunity to go to the school when my company got a slot to fill and I was available. My first free fall jump showed me that if I liked static line parachuting, I was going to love free fall parachuting.

My first free fall jump was a HALO or High Altitude Low Opening jump. We were jumping at Mackall Army Airfield from a CASA-212 aircraft which is like a mini C-130, in a way. We went up to altitude, which for the first jump was going to be at 9000 feet above the ground. That was much higher than the high-mark of 1500 feet I had jumped from with a static line parachute, and the distance to the ground was a little unnerving. Still, I had plenty to think about and was ready for the experience.

Exiting the aircraft was a lot different than with the static line parachute. It was a bit like diving into a pool instead of jumping in feet first. I dived off the ramp and flipped over end to end a couple of times before I got a good solid arch and became stable. Once I was stable, my body seemed to lock into the "flat, dumb, and happy" position and I never had much of a problem getting, and staying stable after that. I was able to do a practice pull and turn a little to line up on the DZ, which was the airfield below.

I opened my chute, which for this jump I had packed myself, by pulling my ripcord handle at 4000 feet above the ground, as was the rule. It opened much differently than my static line parachute opened. I suppose not having a large aircraft pulling the chute out of the bag made it open a little slower. Still, when it did open, it was quite the jolt as I suddenly was setting in a strap saddle at about 3500 feet above the ground. I immediately grabbed hold of the risers as if I could keep myself from possibly falling out of my harness. I felt like I was way too high to be flying around underneath a parachute all by myself.

But it quickly started to grow on me and I fell in love with not only the sensation of flying, but with the amount of time I was able to be under canopy before landing.

Soon enough though, I was approaching the ground and instead of coming basically straight down, I was moving across the airfield at a quick clip. I had turned the chute into the wind, but with not much of a breeze blowing that day, I was still moving forward faster than I was used to going with a round parachute. As I neared the ground, I pulled down on my steering toggles to flair the parachute and slow it down for landing. All went well, and the chute reacted as it was supposed to do, except that I had flared a bit too early and I was still about ten feet in the air when my chute stopped.

I went ahead and made a dynamic PLF when I hit the ground and was fine, although my instructor who had been watching me land, made a note of my poor landing technique. My next jump was better and I now could judge where to flair. On my second MFF jump, I produced a nice standing landing, just like it was supposed to be done. I was hooked and never looked back. My life, as far as parachuting was concerned, had changed and it changed for the better in more ways than one.

Jump 73 a Parachute Malfunction and Reserve Ride

Within days of graduating from Military Free Fall School, I was assigned to the Company's MFF Detachment, SFODA-594 and we deployed to El Centro, CA for some advanced MFF training. This was in February 1988 and all available MFF Detachments in 5th Special Forces Group attended the exercise. The primary operation being practiced during this train-up was the technique known as Stand-off Parachuting, or otherwise known as HAHO.

This type of jump required exiting the aircraft at a high altitude and deploying the parachute while still far above the ground to allow a long movement into a targeted area. The concept was to enable a silent infiltration, as well as one that would generally be invisible to any aircraft radars. During this period of time, a Stand-off jump could be conducted from up to 30,000 feet above sea level and we could deploy our parachutes at any planned altitude, depending on pre-jump planning. Most of our Stand-off jumps had us deploy our parachutes within five-seconds after exiting.

Years later, the regulations were adjusted to require jumpers to fall a minimum amount of time before deploying their parachutes, in order to allow them to slow down to terminal velocity, that is, the speed in which gravity pulls you towards the earth. This helped prevent parachutes from blowing cells apart or breaking lines during the opening sequence and requiring the jumper to cut away, or release the main parachute and then deploy the reserve parachute.

We had been jumping a couple of times a day for several days in the vicinity of Lindsey DZ, near El Centro, CA. We had been exiting the aircraft between 13,000 to 20,000 feet above the ground, and from five to fifteen miles away from the landing site. Usually the first jump would be the higher jump and we would need to use supplemental oxygen during the flight up to jump altitude and during the jump itself. With the oxygen, the rucksacks and load-bearing equipment, weapons, and the parachute and harness, the load was pretty heavy and sometimes getting out of the aircraft posed its own set of problems. Especially in the dark under dim red lights.

This particular jump was in the daytime and we were jumping oxygen on top of all our other equipment from a C-130 via the paratrooper doors. These are the crew doors, the ones on each side of the aircraft towards the rear of the plane. We left the aircraft by turning our bodies just as we got to the door and leaping sideways into the air while facing into the direction of flight, or towards the front of the aircraft, feet downward. This is known as a poised exit.

As I jumped through the doorway, I bumped against the rear edge of the opening and spun myself around as I dropped away from, and behind the plane. This was a Stand-off operation, so I planned to open my chute at the five-second mark to make sure that I was properly located within the parachute stack as we flew towards the LZ under canopy.

The timing or counting sequence we used to keep us all at about the same opening altitude was as follows:

Arch-one thousand, leave the aircraft and remain (or get) stable.

Look-one thousand, look at and visually identify the ripcord grip if it was light enough to see it.

Reach-one thousand, bring your right hand to the ripcord, tracing the cable housing and moving the oxygen hose out of the way if required, and grab hold of the handle.

Pull-one thousand, pull the ripcord and release the main parachute deployment bag.

Clear-one thousand, look over your shoulder to allow the air to catch the deployment bag and begin to open the parachute.

This usually would take about 500 feet if done properly, and it also created some very hard openings at times, based on the speed of the aircraft during exit.

As I've noted, I banged against the doorway as I exited and spun around during my pull sequence. As my parachute deployed and opened, perhaps because of the exit, perhaps because of the way it had been packed, about half of the canopy failed to open and I was starting to spin around in a violent corkscrew motion as I looked upward and observed my canopy.

I began to pulling down and releasing or pumping my risers as I attempted to get my parachute to fully open. When that didn't fix the problem, I grabbed the Velcro attached pillow that was connected to the cut-away or release cables for my main parachute and gave it a hard pull. To my surprise, it only came about halfway loose and my hand slipped off from the resistance. I immediately grabbed it again while stopping my motion with my left hand that was now about to pull the reserve parachute's ripcord. The second try with the cut-away pillow, which probably took about half-a-second, worked and I went back into free fall with the reserve parachute stretching out behind me. It seemed to take a long time for my reserve to inflate. I was no longer falling as quickly after my main opened as far as it did, so as I opened the reserve, I was starting the deployment from having almost slowed down vertically to a standstill.

Because this was a daytime jump, several of my teammates already under canopy could see what had just happened. I was a couple of thousand feet lower than everyone else in the stack so I knew that I would be the first one on the ground and if anyone had noticed, they would follow me to wherever I landed.

This happened to be a Stand-off jump close enough to the LZ that since it was daytime, I could already see where I wanted to land across the desert beneath me. I quickly turned straight to the LZ and managed to fly all the way, landing safely at the intended location. The rest of the Team landed soon after and I was immediately told that I owed the Team a case of beer for having had my first cut-away.

You want to get back on the horse as soon as you fall off, and we were scheduled for a night jump, a repeat of what we had just done, only in the dark, in just a few hours. I was a little nervous as we got ready and during the flight to the release point, I thought about what I would do to improve my technique after the last jump. I thought about it a lot but tried hard not to show any nervousness. I worried for nothing and as it turned out, I had a great exit on jump number 74, and all went exactly as it was supposed to go. The Team once again landed on the LZ and we continued to train.

And from then on, I always removed the cut-away pillow from my parachute and put it back before each jump to make sure that the Velcro connection had not gotten pressed into a super-tight attachment. Fool me once.

Jump 163, 5th Group MFF Exercise 1990

By this time, I'd been on SFODA-594 for almost a couple of years. I'd completed the MFF Jumpmaster Course and made a lot more jumps in the US, and throughout my operational areas in Africa and the Middle East. At this specific time, I was once again deployed with 5th SFG(A) for a group-wide MFF train-up. This time we were in Yuma Proving Ground, AZ and we were able to take advantage of the wide-open spaces that provided a lot of clear, aircraft-free, air space. We conducted a few HALO jumps, but the majority were Stand-off jumps. This jump, number 163, was one that had to be defined as special, simply due to the distance between the RP (Release Point), that is, the point at which we exited the aircraft, and the LZ. I was the primary jumpmaster for this operation and we had two SFODAs on our bird, a C-130. The LZ, the place we were planning to land, was called LA DZ, and it was located on the edge of the reservation in the middle of an ancient sea bed, surrounded by mountains and hundreds of saguaro cacti, most being 20 feet tall, or higher. It was also located 25 nautical miles from the RP.

I had calculated our release point early that evening for the night's missions. There was a very fast jet stream above 15,000 feet and the RP was further away from the LZ than any I had determined before. I ran my figures past not only my assistant jumpmaster, but a couple of others as well. The numbers all added up and I had not gotten my distance calculations wrong, so I felt comfortable with what I'd plotted on the map. We started our flight around midnight and of course, the aircrew wanted to do some low-level flying for an hour, so everyone settled in for a rough ride.

We were only going up to 18,000 feet above the ground, so we didn't need to pre-breath before we got to altitude. Pre-breathing is when the parachutist breathes pure oxygen from tanks located in the aircraft for at least 30 minutes prior to depressurizing the aircraft at jump altitude. This purges the nitrogen from your blood which prevents it from forming bubbles in your blood stream when you go above 18,000 feet. With nitrogen in your blood, a jumper can suffer decompression sickness, or the bends, in much the same way a SCUBA diver can from surfacing too quickly from the depths.

During our ascent to altitude, I rode in the cockpit of the aircraft and monitored the wind speed and direction from the Navigator's onboard computer as we climbed. Comparing these readings to the figures I had already calculated prior to departure, I was able to confirm or adjust my RP with the latest data available.

As the primary jumpmaster, I confirmed the RP with the 1st Detachment's JM and we prepared for the 1st jump run. As it was dark and there was nothing below us except desert, we had sent a vehicle to the RP location to place a light on the ground to aide in guiding the aircraft and assist the jumpmaster with his adjustments. As we lined up on final approach, the spot was verified and the 1st Detachment departed the aircraft.

The plan was to exit at 18,000 feet and to be under canopy no later than 17,000 feet above the ground. I had also built a 2000-foot buffer into the calculation for safety, just in case someone went a little low during opening. The extra altitude could easily be burnt off as the jumper got closer to the LZ.

As we made our 2nd pass about 20-minutes later, I stood on the edge of the ramp and started looking for the RP. During the time between the first pass and my Team's pass, some low-level clouds had moved in and I could no longer see the RP or anything close to its location. The pilot called back to me and asked if I wanted another go-around. I had noticed earlier on the map, that off to my right and perpendicular to the RP, the Colorado River made a sharp double-bend a few kilometers from where the RP had been plotted. I quickly moved to the other side of the ramp and looked out towards the river where I could see it through some holes in the clouds as it reflected the moonlight from above. I quickly adjusted the aircraft to where my best-guess was for the RP below and signaled for my team to stand-by and then about 30 seconds later, to go. Each jumper left the ramp jumping backwards or twisting around to be looking back at the aircraft as it flew away. Using our 5-count method, all deployed their parachutes within the 5-seconds required to maintain common altitudes at opening. I was the last one to leave and as I opened my chute, I could see a few of the strobe lights we had attached to the back of our helmets flashing out to my front as we adjusted our direction of travel with our compasses and got on line with our LZ, 25 miles away as the crow flies.

Being under canopy at night and watching the ground below move away as you flew along in a jet stream was quite an experience. At one point I flew along a stretch of highway and from the vehicle's headlights, I could tell that I was going faster than they were on the road far below. I flew under canopy for over 30 minutes before I reached the LZ and could start to make out the lighted wind direction indicator on the ground. I still had plenty of altitude and started to make slow right and left turns in order to avoid flying past the LZ as I burned off altitude.

Almost at once, like a group of grey ghosts, I started to see the other canopies seemingly materialize out of the darkness. By this time the sun was just starting to come up and it was becoming light enough that I could see the majority of the parachutes all stacked up as we made our final turns to come in and land. I was still the highest parachute, having exited last. Once I had landed we were all accounted for and safe on the ground. The only negative part of the jump was that it was very cold and I was glad that I had mounted my rucksack on the front of my parachute harness as it helped block the wind at altitude. I never had another canopy ride that lasted as long as this one had been, but several came close.

Jump 280, DZ32, Harare, Zimbabwe

During one of my deployments to Zimbabwe, I had the opportunity to jump one of the Zimbabwean chutes that the Zim Air Force's PJI or Parachute Jumping Instructors used during their MFF airborne operations. It was designated as a military parachute and that made it legal for me to use as we were authorized to share each other's equipment if it was a mission requirement. Using one of their chutes might have been a stretch of that logic, but we were deployed overseas and sometimes the rule book got somewhat thinner outside the US.

The US Army's MFF parachute that I was jumping at the time was designated as the MC-4 Ram Air Parachute System. It is about 380 square feet of canopy with both the main and reserve parachutes being almost identical. It could hold a suspended weight of almost 400 pounds, which counting the jumper, the rucksack, weapon, and the parachute itself, was tested all the time at the top end of the scale. The Zimbabwean parachute was a 220 square foot canopy that was manufactured for skydiving. I was about to learn the difference between driving a truck and driving a sports car.

Our Zimbabwean counterparts gave me and two other members of my team the run-down on the Cruise-Lite parachute. It used a small drag chute to deploy the main parachute and it was called a "throw-out" deployment method because you would grab the small chute from its pouch on your leg and throw it out to allow it to catch air and deploy your main parachute. Our American rigs used a chest-mounted ripcord grip, so this would be something much different. We practiced in a training rig on the ground until we were comfortable with that procedure, as well as becoming familiar with the rest of the system's specific details. Soon, we were ready to get up in the air.

The Zimbabwean Air Force was flying the Casa 212 as their primary troop transport aircraft. We could get about 20 personnel into one for most operations. On this jump, however, there were only the three of us Americans and a couple of the PJI jumpers going with us.

The parachute felt more like a small backpack than what I was used to with the MC-4. I didn't know it then, but once I was assigned to the Military Free Fall School back in the US as an instructor, I'd be jumping a similar parachute system a lot. For now though, it felt like I was missing most of my parachute as we waited for the aircraft to climb to 9000 feet. Once we were on jump run and over the DZ, I made a diving exit and could feel the difference in free fall by wearing such a small chute as compared to wearing the larger MC-4. Still, the real difference came as I started getting ready to deploy the main at 3500 feet above the ground.

The smaller parachute being launched by the attached throw-out chute, coupled with the way the main parachute had been folded when it was packed, opened much faster than what I was used to seeing happen. Almost instantly, I was flying, not beneath a large grey-colored square working chute, but a multi-colored, small and very fast, sports chute. I was amazed at how quickly the parachute reacted to my pulls on the steering toggles and I cranked down on each of them at different times, spinning in alternating quick, tight spirals, that made me think of a carnival ride. Soon enough I was close to the ground and needed to line up for my landing. I was able to control the flight much easier than with the larger chute and touched down literally on the center of the marked IP, or impact point, identified on the ground by a large orange X.

I greatly enjoyed the jump and was quite happy with my experience. As I gathered the chute up, I noticed then, just how thin the material felt and how deeply imprinted the fold marks appeared to be in the canopy. I asked the Zim jumpers some questions about the history of the chutes and discovered that they were quite a few years old and had seen many jumps put on them. Probably way too many. Especially considering that there was quite a big weight difference between the average Zim jumper and us Americans. I decided that we shouldn't jump them again and push our luck. But the one jump, that was a good one and I'm glad I tried it out.

Jump 327, MFFS Instructor Training

Getting assigned to the Special Warfare Center as an Instructor for the Military Free Fall School meant that you went back to school to learn how to fly. Not everyone made it. Being a MFF parachutist, even being a MFF Jumpmaster didn't mean that you had to fly. It meant that you could remain stable under almost any circumstance, move in controlled and directed ways, such as moving in a direction of choice, turning towards the drop zone, moving away from someone who got too close, and performing a controlled opening of your parachute at the proper altitude.

Being a qualified Instructor meant that you had to master and pass graded evaluations of your flying skills. Basically, you had to learn how to catch an out-of-control student in the air and stabilize the student in time for a safe deployment of the student's parachute. It meant that you had to move between students in free fall and give each of them visual instructions coaching them on skills required to have a stable and controlled free fall. And you had to do all of that while remembering things like altitude, student numbers, and actions taken or missed, as well as observing the students as they opened their parachutes about 1000 feet before you deployed your own.

There were many air skills the candidate Instructor must learn before he was even allowed to jump with students. One of the skills was called the 20-second drill. For this drill, an Instructor evaluator would exit the aircraft while the candidate Instructor remained inside for five-seconds. At the mark, the candidate would exit the aircraft and had to fly to the Instructor who had left the aircraft five-seconds earlier and establish physical contact and control of his free fall within 20-seconds from the time the candidate jumped out.

On jump 327 I passed my drill for this specific task by completing the process in 17-seconds. It took a couple of tries to get it done, and that was after I had completed other skill tests building up to this one.

Once the individual skills had been mastered, evaluated, and passed, the candidate Instructor had to pass a series of drills by jumping with a qualified Instructor who was in the role of a student having a difficult jump. As the Instructor, I had to not only observe and report what had happened after the jump was over, but I was expected to physically react in the air, and when possible, correct whatever the student was doing wrong that was creating a problem. Remembering all the details was as hard as working with the student in the air. Things happen really fast while in free fall.

Once this phase of the candidate's Instructor training was completed, I was allowed to start working with real students. This also required the candidates to run the show for the entire class session under the supervision of the qualified Instructors and the selected Instructor leader for the session.

The first class was as a "Shadow" Instructor and you followed a qualified Instructor, mirroring all he did and doing what was asked or directed, based on the level of ability you showed during the jumps.

The next class I was given one student per jump to follow and evaluate or assist as necessary. This class included not only graded jumps for the students in the course, but also for the candidate Instructors. I passed my first course evaluations and the next course I was given two-students per jump to evaluate and assist.

During this phase of training I failed one of my final graded evaluations when I had a student with a rucksack tumble out of control and get away from me. I simply couldn't catch him in the air and my evaluator decided that I needed some extra training on my ability to do a faster head-down dive.

This extra training paid off and during my second course with two-students, I passed all my evaluations and became a qualified Instructor. My candidate status ran from June1994 through February 1995 and then I was awarded my Military Free Fall Instructor rating of MFFI-336. I was very proud of earning this rating.

Jump 425 was a 22-Way Blue Angels Show Opener

The US Navy Blue Angels uses NAF El Centro, CA as its winter training base and the MFFS also operated from the same base for several courses during the transition between the MFFS being stationed at Ft Bragg, NC and seeing it moved to YPG, AZ. During the transition period of several months, the MFFS was temporarily conducting most of its jump operations out of NAF El Centro. While working there, the command and operations elements of the MFFS got to know the Blue Angels and the operations & planning department folks pretty well as we deconflicted air space and shared several of the base facilities with the pilots and support crew.

On occasion, the Blue Angels had asked the MFFS Instructors to make a jump as a show opener. One such occasion was when I made jump 425 and we also broke a MFFS record by building a 22-way in freefall.

A 22-way is when 22 jumpers exit an aircraft and physically fly into a prearranged formation, hold on to each other as the formation builds and then hold it until it's time to break apart and move to where everyone can safely deploy their parachutes.

For this airborne operation, we were jumping directly over El Centro NAF from 15,000 feet above the viewing stands. The stands had been set up near the runway that the Blue Angels demonstration team would be flying over during their air show. For the show opening act though, the MFFS Instructors we making the first flight. We jumped from the Blue Angels support aircraft, a C-130 famously known as Fat Albert. The Blue Angels C-130, like the F/A-18 Hornets of the Demo Squadron, was also painted in the famous Blue and Gold color scheme.

My position in the 22-way was one of medium-difficulty for a free fall maneuver, since I was one of the jumpers building the 2nd ring of the formation. The first circle had five jumpers and was built almost immediately upon exit, as it was basically dragged off the ramp as we started leaving the bird. The next ten jumpers had to then fly to their position behind the 1st ring of jumpers and dock with the correct jumper, forming a second ring. The remaining seven jumpers had to then fly behind the correct members of the second ring and build the final points.

The heaviest jumpers usually formed the base and the lighter ones the edges because the larger the formation grew, the slower it began to fall. Without a lot of practice and experience, the last jumpers coming to the formation could easily come in too fast or drop too low and not be able to join the formation.

The formation built very quickly on this particular jump and we held it for several thousand feet above the crowd until we reached our break-away altitude. At the planned altitude, all the jumpers turned away and moved into clear airspace. Chutes started deploying at varying predetermined altitudes, in order to safely disperse the parachutes and allow them to land on the ground in roughly three groups. I was forced to move at the last moment by a jumper who got beneath me as he was opening his parachute. By doing so, I opened below my minimum safe-altitude. It was a quick ride to the ground, but I looked marvelous. I also heard about it later from my safety officer who was also jumping and happened to see me go low.

We landed on the taxiway within feet of the crowd and was able to quickly mix with the folks, many of whom wanted us to autograph pictures of the Blue Angels that had been given out beforehand as part of the show souvenirs. During my time at the MFFS as an Instructor, I made several jumps with both the Blue Angels and the Golden Knights, the US Army Parachute Team.

Jump 475, Bad Student Jump and an Injury Producing Save

This was a typical day jumping with students in a grouping exercise. As the group leader, I poised from the ramp of a C-130 at 12,500 above ground level (AGL) over Lindsey DZ in California. This was the 6th grouping exercise for the students and they were jumping with rucksacks. There normally was a dozen or so in a grouping jump and we would usually have at least three Instructors jumping with the students. I was jumping with the lead students as the group leader and was standing at the edge of the ramp, with my back towards the sky, and holding the chest straps of two of the first students in the group. As we got the GO signal from the jumpmaster, I stepped off backwards, pulling the students behind me as we became the target for the remaining students to try to remain near during the free fall portion of the jump.

During the free fall portion of the grouping exercise, the students would spread out and do a few practice turns, or simply remain stable and fly their position until 4000 feet AGL, when they would pull their ripcord grips and deploy their main parachute. Once open, the students would look for the Instructor chute, usually a dark blue chute, and they would begin to build a stack behind it as they followed the Instructor to the landing spot.

During this jump one of the students who first exited the aircraft became unstable and started to tumble through the sky. When a jumper tumbles, especially with a rucksack, they fall much faster than when they are stable. I chased the out of control jumper several thousand feet and stopped his tumbling, held him in a face-to-earth position and watched him pull his ripcord. I then deployed my chute as I was going to lead the students as they practiced grouping under canopy.

At some point during my struggles with the tumbling student jumper or during my own opening shock as I deployed my parachute, I strained a few muscles in my back between my shoulder blades and the base of my neck. By the time I had reached the ground, I could barely lift my right arm. I actually had to pull myself off the next scheduled jump and was unable to jump for the next week.

The act of catching a jumper that is tumbling or spinning out of control is not an easy task and you can easily become injured by being struck or wrestling with the centrifugal force created by the mass and speed of the jumper. I later found out that I had twisted a couple of vertebrae slightly out of alignment which had caused the initial pain. After a week of rest and a lot of Motrin, the pain went away and I regained my arm functions, but from then on, I would occasionally have the same injury flare up and cause problems.

Jump 583, Stadium Jump

It was October 28, 1995 when I made a Stadium Jump for a college football game opener at the Arkansas State University in Jonesboro, AR. I don't recall why the MFFS was asked to make the jump, or by whom, but I know I was glad for the chance to get to do something so different. The jump team was to be me, the MFFS Commander, Major Wes Rehorn (Cowboy 6), and one of the retired SF Instructors still working as a DOD Civilian. Also, the SWC School House Operations Officer, retired CSM Dave Clark, as well as the BN's senior NCO, CSM Griffon. Of the five of us, I would be carrying the game ball, and the Major would be jumping in the American flag.

We flew there in a military Twin Otter early in the morning and made a practice jump. The stadium is like many others and built in a large hole with the stands all around creating a giant bowl-shaped DZ.

For the practice jump, all went well and we landed on the field with just a little turbulence from the wind that was blowing around the very high, stadium superstructure on one side of the field. The top of the structure was easily a hundred feet or more above the playing field. The wind was within tolerance, but it was hitting the upper limit for jumping, especially into such a tight and obstacle-surrounded space.

A few hours later it was time to make the game jump and the wind had not let up and had actually gotten worse. We were told during jump run that the winds were blowing between 15 to 20 knots, which was right at the cut-off point. The pilot asked if we were going to jump or stay inside and land and we decided to make the jump. I found out later that once we left the aircraft and had started our jump, the wind had picked up to 28 knots. As soon as we opened our parachutes, it became painfully obvious that the wind was blowing much stronger than was safe.

As we started to set up for our landings, two of the last jumpers to leave the plane turned away from the stadium and landed in an open area away from the stadium's parking lots. I attempted to steer towards the football field and was being blown backwards the entire way. Once I determined that I would not be able to get inside the stadium and make it to the field, I turned towards the closest parking lot and started to land on the roadway between the parked cars. Just as I got close to the ground, a gust of wind blew me about ten feet back up into the air and over into the line of vehicles, where I landed on the front fender of a U-Haul rental truck. I wasn't hurt, but I did put a dent in the truck's fender. I packed my chute into its deployment bag and quickly trotted into the stadium with the game ball. I entered the stadium through the stands by one of the touchdown zones and no one even noticed that I had landed in the parking lot.

While I was landing, I saw that the CO had managed to get into the stadium field while flying the Stars and Stripes beneath him. He had just cleared the stadium seats when the wind slammed him down really hard and while he stuck the landing, he severely twisted his ankle on impact which kept him from jumping again for several weeks. He was the only one who actually made it into the field for the second jump, but as he landed I showed up on the field with the game ball and the crowd thought we had both landed there. We got a lot of applause and cheers, and about 5-minutes of fame. The flight back to the MFFS was filled with talk of how dumb we had been to make that second jump. It really was pretty crazy but it also made the local evening news. We were famous!

Jumps 617-628 Tandem Master Training, MT#17

In February 1996, I had the opportunity to take the Tandem Master training course that the MFFS headquarters element had just completed designing and was going to run for the very first iteration with a few MFFS Instructors as the students. There were very few military tandem masters and most were in Special Mission Units. My training consisted of eleven tandem jumps ranging from first jumping as a passenger, to jumping with an extra heavy rucksack instead of a live passenger, and then jumping with real people as passengers for the rest of the training. The jumps included jumping with just the passenger and parachute, to combat equipment (rucksacks and weapons for each), oxygen systems for each, and all items at once for each jumper, both day and night jumps. Even as a seasoned MFF Instructor, this was a quite different and stressful experience.

Initially, the parachute system itself was designed differently, to include even the ripcords. Also, the emergency procedures were different. The parachute itself was much larger than the normal MFF parachute as it had to hold so much more weight. For a couple of the jumps, the passenger who was also an experienced tandem jumper for these specific jumps, would create problems in free fall such as a panicked passenger might do on a jump. The tandem master student had to correct the issue, and that could even mean punching the passenger in the head, or holding his arms or legs with your own.

My jump number 626 was unique because it was being videoed and was being captured via still photographs by one of our MFFS Instructor photographers. It was one of the photos taken during this jump that ended up being used as my going-away plaque when I was reassigned back to a Special Forces Group at the end of my school house tour of duty. Also, one of the photos was framed and hung at the MFFS HQ building in YPG.

During this photo-shoot jump my passenger was a MFFS Rigger, SGT Eatmon. Eatmon was a large lad and we both wore rucksacks and rifles as combat equipment, plus we each wore oxygen systems. With this equipment and the parachute system, along with our combined weights, I had calculated that the total was about 630 pounds. Being fastened together and weighing this much made moving around in the aircraft difficult. I was also doing my own jumpmaster spot for this jump and that was a challenge as well. Still, it went fine and soon we were outside and in free fall. We made a diving exit and I was able to "set the drogue" without any problems. Because of the built-in instability of jumping the tandem system, the free fall was usually done with a small drag chute that helped keep the jumpers stable until it was time to deploy the main parachute. During the free fall we purposely made a complete right and left hand turn and I made a couple of practice pulls for the deployment sequence. Once under canopy, the flight continued to go well and we landed right in the pea-pit on Phillips DZ, YPG, where we did all our jumps for the tandem course.

The next jump was a bit different.

Jump 627 was a repeat of jump 626 with the addition of being conducted at night. One of the differences between the normal MFF jump and the tandem jump comes at the exit. If for some reason you can't get stable, you won't try to set the drogue chute. You don't want to risk getting the lanyard wrapped up if you are spinning or tumbling. Also, if you can't get stable, you must quickly decide to activate the reserve parachute instead of the main. Again, if the main was deployed and became entangled with the jumpers or the equipment, you would have a malfunction that would probably kill both jumpers. The reserve is made to deploy and not become entangled with the deployment bag and equipment or personnel.

So, we made a diving exit and immediately started spinning to the right. I countered and stopped the spin, but then we went head-down and began to tumble. After about two tumbles, I stopped it, but we went right back into a hard-right spin again. Being dark and wearing the oxygen mask prevented me from seeing much of what was happening with my passenger, but later I recalled seeing his altimeter light as it was waving around. I assume that he was flailing around and creating the instability problems instead of remaining in his normal passenger position. Regardless of why the jump was starting off so FUBAR, I made the decision that this was now an emergency procedures event and I went for the reserve parachute deployment handle.

We were safely under canopy at 7000 AGL and I began to burn altitude in order to land on the DZ, which was almost directly beneath us. After what seemed like a really long time, we finally landed and hit the intended landing spot right on the tip of the lighted wind indicator arrow laid out on the ground. Due to the long time it took for me to get down to the ground, several of the DZ party had started to think that we may have landed away from the DZ or worse, had a malfunction. They were getting ready to start a ground search. The DZ Safety Officer and crew were quite glad to see us as we got nearer the ground just before landing.

The very next day I was hitting the silk again and with the same equipment as the night before. My passenger this time was a non-MFF qualified soldier who wanted to experience the ride. Once more, as before the events of the last night's jump, everything went picture perfect from exit, to drogue-set, to free fall, to deployment, to landing. This was the final graded evaluation and having successfully completed the jump and the program, I became the military's 17th official Tandem Master.

Jump 663, My Final Parachute Jump

My final parachute jump, # 633, was a nighttime jump from a C-141 cargo jet with all the combat equipment we could strap on, and I was once again, the Jumpmaster. We jumped over an airfield in Indian Springs, NV from 1500 feet AGL. Not knowing at the time that this would be my final jump, I missed the irony of how it was also being provided by the same platform, a C-141, as my first jump at Ft Benning, GA, in what seemed like a lifetime before.

However, fate was not all smiles as there was soon to be a fly in the mix, a bump in the road, or as typical during many Airborne Operations, a USAF wrench in the process. About the time the first pass had been given the “Stand Up” jump command, I was notified that the scheduled number of passes, or jump runs, had been reduced from five to three.

After a two-hour low-level flight over the Rocky Mountains (as low as possible), flying between mountain peaks and down valleys, suddenly I had to quickly start doing mental-math to not only decide how to rearrange the jumpers into three passes, but also how to adjust the assistant Jumpmasters in order to maintain currency in some of the ones who were getting close to becoming non-current. It was a good thing my initial SF job was as an Engineer, as we had to be pretty good at math. Without much confusion though, the sticks were quickly adjusted and the proper amounts of jumpers were able to exit the aircraft in three passes, emptying the stripped-out version of an airliner of all the no longer required passengers.

I landed under my canopy near a big wind-indicator between a runway and taxiway and soon found out that all the jumpers had been accounted for and that there had been no reported injuries. My jumping days were over and my record was good. Airborne!

My Green Beret Trek

In October of 1986, I climbed aboard a 5-ton truck and headed from Fort Bragg, NC for an almost 45-mile ride to Camp Mackall, NC. I was setting in the back of that truck with a group of others and all of the gear we could pack in our large ruck sacks. Our truck was only one of a few that were also filled with a bunch of soldiers who were also starting a trek that upon successful completion would not only change our jobs in the Army, but change our lives forever. Just as we got near the gate the trucks stopped and we were told to dismount. Now that we were entering the place we were to call home, we were going to walk at a very fast pace while carrying everything we owned. We knew it had started and the first gut-check was about to begin. Everyone in my class was hoping to change from our current specialty and become Special Forces branched soldiers. I was going to leave my 13 Series MOS (Artillery) and strive hard to earn an 18 Series MOS (Special Forces) as a new Green Beret wearing Special Forces operator.

My SF training would be focused on the 18C, or Engineer Sergeant track. This MOS was one of eight, counting the Officer and Warrant Officer, that would fill out a fully-manned SF Operational Detachment- Alpha, or SFODA. The Special Forces Engineer Sergeant employs offensive and defensive combat engineer capabilities; performs and instructs in demolitions, explosives, improvised munitions, US and foreign landmines, mine and countermine operations, construction, field fortifications, bridging, rigging, electrical wiring, reconnaissance, target analysis, and civil action projects. He also instructs and performs land and water navigation duties by interpreting maps, overlays, photos, charts and using standard and nonstandard navigational techniques and equipment; is proficient in sabotage operations with standard, nonstandard, and improvised munitions and explosives; plans, prepares and conducts the target analysis portion of the area study; collects and disseminates engineer intelligence data; prepares and reviews target folders; and prepares and reviews target analysis folders.

These skills, along with what were called common skills, were expected of the 18C. There were two 18C slots on an A-Team, one rated as SSG (E6) and one as SFC (E7).

The common skills consisted of many Special Operations focused requirements and included a demonstrated tactical competence in areas such as Unconventional Warfare, small unit tactics at the squad and platoon level (as a starting point), SERE or Survival, Evasion, Resistance, and Escape missions, Advanced Spec-Ops techniques, Air and Water operations, military decision-making skills, as well as all the combat-related skills of the Infantryman's shoot, move, and communication requirements.

At the time I went to the Special Forces Qualification Course, you had to have been in the Army long enough to become a Specialist (E4) or higher. SF wanted a more experienced and mature candidate. After 9.11.2001 and the Global War on Terror (GWOT) started, SF needed more bodies and opened the field to all, regardless of rank or experience.

I've outlined the highlights of my Special Forces career below and I will detail some of the specific events which can be shared. I believe the reader will find this interesting and insightful.

SPECIAL FORCES ASSIGNMENTS:
5th Special Forces Group (Airborne) Fort Bragg, NC
3rd Special Forces Group (Airborne) Fort Bragg, NC
1st Special Warfare Training Group (Airborne) Fort Bragg, NC,YPG, AZ
10th Special Forces Group (Airborne) Fort Carson, CO

The Ballad

No Special Forces history page would be complete without Barry Sadler's Ballad of the Green Berets. I grew up listening to this as a young boy, never thinking that I would earn my own Green Beret one day. Written by SSG Barry Sadler, 1965

Ballad Of The Green Berets

Fighting soldiers from the sky
Fearless men who jump and die
Men who mean just what they say
The brave men of the Green Beret

Silver wings upon their chest
These are men, America's best
One hundred men we'll test today
But only three win the Green Beret

Trained to live, off nature's land
Trained in combat, hand to hand
Men who fight by night and day
Courage deep, from the Green Beret

Silver wings upon their chest
These are men, America's best
One hundred men we'll test today
But only three win the Green Beret

Back at home a young wife waits
Her Green Beret has met his fate
He has died for those oppressed
Leaving her this last request

Put silver wings on my son's chest
Make him one of America's best
He'll be a man they'll test one day
Have him win the Green Beret

SFQC, The Hardest Thing I've Ever Done

Phase 1

When I went to the Special Forces school, I was doing so not for a love of becoming a Green Beret, but because of (as they say) a series of unfortunate events. I had attended Jump School at Ft Benning after my assignment in Germany from 1983 – 1985. Josh had been born in Germany and now we were all assigned to Ft Bragg, NC. I was assigned to the 18th Airborne Corps Artillery and was working as a newly promoted Staff Sergeant (SSG E6) in the Tacfire section as a Chief Computer. I used the Tacfire computer system to plan and control artillery support for the elements of the 82nd Airborne Division.

Once Artillery had transformed from the old-school "charts and darts" or manual fire support, to the computer-driven Tacfire system, the job of the Fire Direction Control (FDC) sections became much more technical and it seemed to me, further from the action I wanted to experience from a Combat Arms job. I was coming up on my time to reenlist and I went to the reenlistment NCO with a plan. I no longer cared for my current job in the Artillery, but I had been an Artilleryman for over six years and was very good at it. I wanted to reenlist as a Forward Observer and for Ranger School. I knew that this option would take me out of the FDC job and place me with the Infantry units on the forward edge of the battlefield, but I'd still be an Artilleryman and would be the one calling FDC for fire missions. The re-up NCO said I couldn't do it because my FDC job was currently at a critical shortage in personnel. I was given two options besides my current job, that I could pursue for my reenlistment. The options were hard hat diver (still in the Army) or as a Special Forces Engineer 18C. Apparently, both of those jobs were more critically short in qualified personnel than the FDC job I currently held. I decided that the Special Forces job was the closest to being an FO in the Rangers, so I raised my hand and signed the dotted line for six more years and a change of scenery, although I was staying at Ft Bragg.

In 1986, there was no pre-SF training selection and assessment course. I got my assignment and became a member of the USAJFKSWCS (John F Kennedy Special Warfare Center & School) as a student attending the SFQC. For me, this was a three-phase course consisting of Phase 1, a small-unit tactics and operations school, not unlike Ranger School. In fact, the Ranger Handbook was the Bible for the entire Phase 1 exercise and the 5-paragraph operations order became a part of everyone's lexicon. The candidates were broken into teams representing SFODAs and each member rotated through the leadership positions a minimum of twice. It was important to not only do well during your time in command, but also when you were not in charge. At the end of the training exercise the members of the student Teams were able to evaluate each of their peers, and a low score for someone who only worked hard when the SF Instructors were watching them could get a failing score during the Peer Evaluation at the end.

Not everyone attending the SFQC had come from a Combat Arms job before reaching the school. Of the Combat Arms soldiers, not all were Infantrymen. The Infantry and Special Forces tasks presented during the Phase 1 missions were quite challenging, especially to those who had little or no experience in Infantry operations. Many of the soldiers needed some extra mentoring along the way. However, the greatest challenge to the Phase 1 candidate wasn't the combat-oriented patrols, it was the individual Land Navigation requirements that took place throughout the training cycle.

Most of the soldiers who dropped out of training or failed and were sent back to their old units did so because of the Land Navigation requirements. The ability to read a map, plot a route, understand how to follow and use a compass, and to do so under great stress, in all weather conditions, and in the dark of night was paramount. There were multiple day and night events and all were graded and all had to be passed to be able to continue training. You could fail a couple of times on the day or night Land-Nav exam, but had to pass them both by the final attempt.

This was not simply like following a road map. You had to move cross-country and if you were caught on a road by the cadre, you failed. You also had to carry your equipment, which weighed in around 75 pounds. And, you had to find multiple points and return to basecamp in an allotted time which meant that you had to travel quickly between the four to five points, most of which were four to five miles apart (or more). These targets were not buildings or easily identifiable landmarks. They were literally metal stakes, the T-shaped fence post type of stake, with a unique identifier chained to them for marking your score card. Sometimes these metal stakes were inside large clumps of brush or otherwise located so that you had to be within a few feet of the stake in order to find it.

My first graded nighttime Land-Nav exam was tough. One of my points was almost impossible to find. In fact, as I got to where I thought the point should be located and searched unsuccessfully, I began to think I wouldn't be able to finish the event. Thankfully, or maybe it made it worse, this was my last point I needed to find.

I had searched without finding it for almost an hour. I had pulled back to a known point and went in from a different attack angle several times with no luck. During this time, several others had come to this point and it seemed that no one was able to find the point. Of course, there was no contact allowed, so all were avoiding each other, for the most part. Suddenly a light flashed a few times and then disappeared. Slowly the personnel hunting for the stake started to move towards where the light had flashed, still avoiding each other as to not violate the honor code. Near where the light had flashed I finally found a green stake about two-feet high, hidden inside a clump of scrub oak brush in a draw, a deep natural cut through the terrain created by run-off.

I felt that the location was really unfair, but I didn't have time to think about it long. I had to get back to Basecamp before my time ran out and I had a long way to go. I followed my route and maintained my bearing, checking my compass and counting my paces as I followed the terrain and made checks of my location on the map. All at once I reached an area that I recognized from where I knew my way back to Basecamp. I had been there earlier in the week. I put my compass into its case, the map back into my pocket, and headed off at a trot for the Basecamp. I soon stepped into a hole left from where a tree stump had rotted away and I fell down beneath my ruck sack and unbeknownst at the time by me, I had also lost my sense of which way I was going. Within a few minutes after starting off again, I sickeningly realized that I had no idea of where I was or which way I needed to go to find the Basecamp. If only I had checked my compass after I had fallen, but self-pity was not helping me find myself again. I started walking in large circles, widening the distance each time around until I found a landmark that I recognized. It was a trail intersection I'd seen before and suddenly I knew exactly where I was and started to run as fast as I could back to Basecamp.

I arrived at the Cadre's check-in point with my score card completed with the stake marker identifiers, but I was literally three minutes late. I explained my dilemma and the SF qualified Cadre listened and kindly told me that since I did so well this night, I'd have no trouble doing it again during the next night. I was terribly disappointed and frustrated beyond belief, but there was no recourse except to move out sharply, eat some chow, and get some rest. Tomorrow would come fast and nothing would get any easier. I passed the next night's event without any problems, but many of my fellow candidates didn't. I was glad to get through the Land Navigation exams and move ahead with training. I also discovered that I would be using my Land-Nav skills a lot in not only my training ahead, but in the real world once I completed training and got to a real Team at Group. Obviously, this was before everyone started using GPS based equipment and moved away from the map and compass. One day, this will come back to bite someone.

Phase 2

After finishing the very physically challenging Phase 1, where we all mastered the common skills and basic fundamentals of the Special Forces small unit tactics and survival techniques required for many Unconventional Warfare scenarios, we were separated by our designated job titles and began to learn our specific skill set for the position we would fill on a Special Forces Operational Detachment Alpha, the SFODA. My job title being 18C, the Special Forces Engineer. My specialty would be construction and demolitions, as well as associated target analysis and planning for combat or military assistance missions. Our Phase 2 classes were comprised of both classroom studies and practical exercises and Engineering-based patrols during field operations.

The most difficult portion of the training, at least for my class, turned out to be the portion that taught us how to classify and build bridges. We would have to classify all types of bridges, from the simple wooden bridge to a multiple-lane highway bridge. Classification of a bridge meant figuring out what type of traffic a bridge could handle without collapsing. We also learned how to build various military bridges, usually from prefabricated bridging sets. Another bridge we learned to design and build was the simple Timber Trestle bridge. This was a bridge made of wood, usually one-lane, capable of handling most military vehicles, and could usually be constructed with materials native to the area. During our field exercise covering bridging, we actually built a bridge like this in about two days. The part of Bridging that made it so difficult and caused the most drop-outs, was the math required to complete the classification and design requirements. Algebra and geometry were put to use in a large way during bridge design.

My favorite part of Phase 2 was by far, the demolitions and explosives training. I was able to learn and practice many different techniques, using many different types of explosives. We became proficient in using various types of demolitions for steel-cutting, cratering or blasting big holes to create obstacles, cutting down trees and poles, and even emplacing demolition ambushes.

During one trip to the demolition range, several of us were selected to make remotely fired weapons for a vehicle ambush using the M72 LAW (Light Antiarmor Weapon). This was usually a shoulder fired weapon, but we were going to rig a wireless, remote-controlled electric firing system to the launcher that would allow it to be fired from a safe distance from the ambush site. There were several systems and I got mine set up and had positioned it between a tree and some rocks. I used mud to get it secured into place and made sure it was properly camouflaged and aimed at the spot I wanted to hit with the rocket, once I fired it from the bunker near the entrance to the range.

Once everyone had finished and we went to the bunker, we all fired the rockets at the same time, each going towards different targets in the ambush zone. My rocket was the only one that hit the target and I was quite happy to have earned those particular bragging rights. Our instructor was a quite famous, retired Special Forces Command Sergeant Major. Ernie Tabata was a Korean and Vietnam war Veteran and taught SF Engineers for decades. Prior to when he died in 2015, it is said every SF 18C had been trained by Ernie. Needless to say, I was glad to make him smile by achieving this small victory of hitting my target with the rocket I had set up. To my surprise and great honor, Ernie presented me with a hard-boiled egg from his lunch box as an award. I was beaming with pride and that was the best hard-boiled egg I had ever eaten.

Another time during my Phase 2 training was equally memorable, if for different reasons. This was towards the end of Phase 2 and we had deployed from our classrooms in the SF training area at Ft Bragg known as Smoke Bomb Hill, back down range to Camp Mackall. We were going through our Field Training Exercise (FTX) as a culmination event that would combine all of our newly learned Engineering skills, as well as the small unit patrolling techniques we had completed during Phase 1.

Once more, we rotated everyone through the leadership positions as each of the 18C candidates took a turn in the barrel to plan the Engineering requirements for the assigned patrol. My mission was to destroy a two-lane Timber Trestle Bridge crossing a large stream named Drowning Creek. Of course, there would be "enemy" soldiers guarding the bridge, so the bridge attack would be more than just putting charges on it and blowing it up.

Being a training mission on a functional bridge, the charges were not made of real explosives, but I still had to use the proper amount of clay, just like it was actual C4 plastic explosives. I also had to construct the charges based on the information I had to collect during our reconnaissance of the target. In addition to the demolitions portion, the assault, support, and security requirements also had to be planned and put into action during the attack. I had decided that we would move covertly to the target late at night, and that I would quietly as possible, slip under the bridge by moving through the water from downstream, plant the explosives on the supports beneath the bridge and then attempt to slip away prior to detonating the demo.

The remainder of the patrol members would be set up in position to attack, should I be discovered by any of the guards. Everything seemed to be working just fine other than being a bit cold, and I was able to move through the water without being noticed by the guards. However, the guards were not the only living things at the bridge. As I made my way under the bridge, towards the trestle rising out of the water in the middle of the stream, in water that was almost up to my neck, with my hands full of demolitions ready to attach to the support beams, I noticed two or three long black things sliding off the bank through the dim red light of my small headlamp. I knew at once that they were water moccasins. North Carolina has a lot of snakes and some of those snakes are poison cottonmouth water moccasins, and I was sure that a few of them were at that moment, headed right for me.

I stopped momentarily, and for what seemed like a couple of minutes I hesitated about what to do. I knew the exercise was pretend, but I also knew the snakes were real and they didn't follow any rules. But, I went ahead and kept going to the wooden trestle and fastened my charges to the support beams just like I had planned to do. I figured if I was to get bit, I'd get bit whether I put the charges on the target or not, and I wanted to pass this graded event.

As I was departing, one of the Cadre must have let the guards know that we were there because I'm sure that I wasn't making any noise. Suddenly the grenade simulators were going off all around and making ear-splitting booms and bangs. The M60 machine gun was blasting from the bridge deck and my Patrol members were firing back from all around. I just kept going back up the creek and met up with everyone as we all started showing up at the predetermined rally point.

The mission resulted in a passing grade of "go" which was 100% better than a "no go" would have been. I told everyone about the snakes, but I'm not sure everyone believed it. Perhaps they weren't cottonmouth water moccasins, but in my mind, they were as bad as the devil himself.

Still, even with the snakes, the mission was fun and the entire culmination exercise was a great test of what we had learned so far, as it continued to build the knowledge base we would all need to be effective members of our Teams once training was completed and we were assigned to real Detachments. Next stop, Phase 3.

Phase 3 (Part 1)

The final phase of the SFQC was where the big difference between being in the Regular Army and the Special Operations Forces really started to come into focus. Up to this point it was pretty much Infantry and your basic requirements of the specific job that you were to hold once you became SF qualified. For me, I could leave right now and work as either an Infantryman or as a Combat Engineer. The Phase 3 element was the unconventional warfare element that made the Green Beret wearing soldiers worthy of the "Special" part of Special Forces.

We began our classroom work back at Camp Mackall, or Camp Rowe as it was to be renamed later on upon the death of COL Nick Rowe. During his time in the Army he had been captured and held captive by the Viet Cong for just over five years before he escaped. He later established and commanded the Survival, Evasion, Resistance, and Escape (SERE) training school which is now a requirement as part of the SFQC. COL Rowe was the guest speaker at my graduation ceremony when I completed the SFQC.

After the classes had been completed for the Phase 3 training in Unconventional Warfare (UW), Guerrilla Warfare, resistance and insurgency operations, specific UW oriented air and other infiltration techniques, as well as a lot of diplomatic operations training, we prepared for the culmination field training exercise called Robin Sage.

The exercise began with the planning and mission preparation known as Isolation. Here every detail was planned and memorized by the Detachment. My position for the exercise was going to be the Senior Engineer and as such, I was responsible for not only the Engineer functions of the primary mission, I was also responsible for all the S4 or Supply and Logistics requirements.

After several days our Detachment provided the Battalion Commander with a mission briefing known as the Brief-back. In a real-world operation you could have several Detachments planning for the same mission and the one that got selected based on their mission brief would deploy while the others either remained in reserve or were given new mission packets.

Our mission required us to insert into the Area of Operations via a static line parachute drop. We flew into the area known as the fictional country of "Pineland" comprised of around 4500 square miles of the Uwharrie National Forest centered on Robbins, NC. The exercise is named Robin Sage and derived from Robbins, NC and COL Sage, a WWII and Office of Strategic Services, or OSS Veteran. The basic mission is to infiltrate a denied area, link up with a Guerilla Force and train them for, and participate on various operations. The exercise requires interaction with not only the Guerrilla Force, but many levels of support, communications, and auxiliary groups. Many of these are made up by local civilians, law enforcement, as well as other units within the military.

My Detachment was to jump in and we had been flying in a C-130 Hercules aircraft for a couple of hours. It was near midnight when we were told to get ready for the jump. As the ramp opened, I could see clouds on the horizon and we were told that the winds on the Drop Zone (DZ) were near the cut-off of 13 knots. We just wanted to get out and get started. We got the Stand Bye command and then the Green Light to go. I was the last man in the line, or stick, as we headed out the back of the aircraft and off the end of the ramp. I watched the Green Light turn to Red as I stepped off the ramp into the darkness and blast of the wind as I waited those few seconds for my parachute to catch the air and inflate.

I could see a few of the parachutes as I descended towards the ground, but they quickly disappeared from sight. As I got closer to the ground I could tell I was being blown backward across the field at an alarming rate and prepared to hit the ground a bit harder than normal. The wind had obviously picked up and I later found out that I had been the last man that would jump that night. The rest of the teams making airborne infiltrations flew back to the airfield and had to come back in trucks. I should have been so lucky.

My rucksack had been lowered on its lowering line and was hanging about 12 feet below me as it hit the ground, was snatched back up as I blew backwards and hit the ground again, just as I landed on top of it. The ruck forced my legs apart and when I hit the ground with my full weight, I felt a terrible pain in my right leg like I had never felt before. I released a canopy connector on my parachute harness and allowed my chute to collapse as I lay on the ground holding my leg and waiting for the pain to subside. After a minute or two, I stood up and started to gather my parachute lines and canopy. That's when my foot twisted sideways and I feel down the first time.

I was pretty sure that I had broken my leg when I hit the ground. I managed to get all my gear and start walking toward the assembly point. About every couple hundred meters, my foot would again twist sideways and I would drop to my knee. The pain was not pleasant, to say the least. Finally, I reached the assembly area and besides the Cadre member who was there to evaluate our mission, there were a few more of my Teammates already positioned around for security. One of them was the medic. The medic and the cadre evaluator checked my leg out and agreed that it was likely broken. By now, I had had my boot off and back on twice. My lower leg and ankle were already black and swelling badly.

As everyone assembled, one more teammate showed up with what he claimed was a broken leg as well. Finally, another teammate had to be located on the DZ as he had also broken a leg upon hitting the ground. As it turned out, there were one broken leg and two badly twisted ankles as a result of that one jump before the airborne operations had been cancelled.

We all earned an ambulance ride back to the Ft Bragg hospital for our efforts. The sprained ankles were able to go back and rejoin the Detachment the next morning. Me, I went into surgery and had three screws put into my fibula. They are still there to this day. My Phase 3, attempt number one, was over. Eight-weeks later, I would be back at it again, a little worse for wear and tear.

Phase 3 (Part 2)

The ride from the drop zone in Pineland to the hospital back at Ft Bragg was a long and quiet ride. I kept thinking how and why did this have to happen to me, and why now of all times? I was so close to the finish line. I was at the top of my Engineer class and if I hadn't messed up royally during the field exercise, I would have been the Honor Graduate for the 18C Course. Would-a, could-a, should-a. But instead, here I was in the ambulance with what I hoped was a sprained ankle, but knew was much worse. When I got to the hospital and made sure someone from the school house had my rifle and gear, I was taken by some medic to the x-ray room. He had taken the other two already. When he helped me off the table and back into the wheel chair he told me that I was the only one with a real broken leg.

I remember I was taken to a room with several beds lined up, more like an open bay barracks room than a hospital room. Once I got into the bed a nurse came by and said she had a pain shot for me. I don't know what it was, Demerol I think, but it burnt like the blue blazes when she injected it into my hip. It actually hurt worse than my leg did. She came back again and said she had another shot for me if I needed it and I passed on it. I was more comfortable with the pain of my broken fibula than I was with what I experienced from that first shot.

She told me that I had a spiral fracture of my fibula and that in the morning I would be going into surgery to receive an internal fixation. I found out later that an internal fixation meant that they were going to cut my leg open and screw three metal screws into my fibula to put it back together. All I could think about was how long would it take me to heal back up so I could finish Phase 3. As it turned out, it would take eight-weeks and a day or two.

I made Linda miserable for a while once I got out of the hospital. The doctor had put a big cast on my leg from my knee down to my toes and somehow wrapped a wrinkle across my shin and after a couple of days it was hurting worse than my ankle did where they cut me open and literally screwed me back together. I ended up breaking the cast up to relieve the pressure on my shin before I went back for my first check-up. I left with a new cast, but this time it was a walking cast and was a lot more comfortable. With my crutches and the big rubber heel on the cast, I could get around pretty well after a few more days.

Meanwhile, one of our neighbors decided she would make my predicament a little better. She brought me a small bell to ring whenever I needed something from Linda. The bell didn't last long. I slowly accepted what had happened as I began to heal. I was given 30-days of medical leave (vacation) and then returned to the school house to see what was in store for me.

I found out that the next Phase 3 class was starting in about three-weeks. I told the Company 1SG that I would be cleared and ready to go into Isolation with one of the student Detachments. This is when I started playing the doctor at the hospital against the one at the Troop Medical Clinic (TMC) at Camp Mackall.

The surgeon that fixed my leg told me that I could get my cast off at about five-weeks and he would allow me to go to the field if I could walk and avoid strenuous activities. I took that information and carried it back to the TMC doc, but adjusted the remarks to proclaim that I would be cleared for duty, any duty, by the time we went to the field after Isolation. He didn't agree, but said he wasn't going to argue against what the surgeon said. Lucky for me, he didn't check out my story.

I had been told by the Company Commander in charge of the Phase 3 training that even though I was cleared for duty, he didn't want me making a parachute jump on my newly mended leg. He stated that with the six or so Detachments that were going into Isolation and then deploying to Robin Sage, only two would be jumping. The other four would insert from helicopter or small boat. He didn't know what my Detachment would be doing, but he said if it was a parachute insertion, then I could meet them on the DZ since I had already made an Infiltration parachute jump during the last class.

When Isolation started, I was placed on one of the Detachments. As it turned out, this class had no Engineer course students, so I was the only 18C (Engineer trained) student in the Company. One of the two cadre assigned to our Detachment was also an 18C and had served on the Engineer Committee prior to moving to the Phase 3 committee. He now decided to make almost every mission an engineer or demolitions mission. Additionally, since this was my second Isolation, I suddenly found myself put in the position as the Team Sergeant. My experience eight-weeks prior gave me a lot of insight into what was coming and what was being looked for during the evaluation of the Isolation portion of the exercise.

My leg, while functional, was not yet at 100%. I was eating what we called "footballs", the 800mg Motrin tablets, like they were candy. But, I was managing. When we got our mission brief, I found out that we were one of the two teams that was to parachute into our Area of Operations. The same DZ where I had broken my leg during the last class. And now, for whatever reason, the Company Commander had changed his mind and told me that if I couldn't jump in with the Detachment, I couldn't stay in this iteration of Phase 3. Of course, I told him that I had been cleared to jump by my surgeon prior to coming out to Camp Mackall for Isolation. He accepted my information and we drove on.

My Detachment Medic wrapped my leg with ace bandages until I looked like a mummy during our prejump training and preparation. Knowing that I had not been cleared for Airborne Operations and getting ready to make another night jump on the same DZ only eight-weeks from when I broke my leg made me wonder if I was doing the right thing. And then I pushed it out of my head and forgot about it. I didn't think much about it again until I was headed out the door of the C-130 roaring about 1000 feet over the same field I'd seen beneath me the last time I jumped from an aircraft while in flight. All the way down I kept thinking "feet and knees together" and to not lift up my sore leg and try to land on my good one alone. Just before I hit the ground, I picked up my sore leg and landed on my good one alone, but nothing broke. I was ecstatic and glad to be alive and as far as I was concerned, good to go. I gathered my gear and headed to the assembly point. The cadre gave us a post-jump brief and changed everyone's duty position. I was no longer the Team Sergeant and much of the burden from earlier was gone. I took a couple more footballs and thought that the hard part was done. From here on, it would be smooth sailing and just one patrol after another with rest in-between patrols. Easy breezy, or so I thought.

A Special Forces ruck sack is always heavy. It's really heavy when one leg only works at about half strength and about 80% of the time. I did well except when the terrain or a small rock would put my right foot at a wrong angle and then between the pain and the weight of all my gear, down I would go, usually to my knee, sometimes all the way to the ground in a heap. The first time or two caused some attention, some unwanted from the cadre, but soon after, no one noticed or cared. We were all tired and everyone was sore in one way or another. I just kept eating Motrin until my ears would start ringing and then I'd maintain that level of medication. I'd worry about my kidneys or liver later.

The next morning, after we had made a long infiltration patrol through the countryside, after we had conducted Stand To, a period of watching for anything bad as the night turns into day, I was sent out to a security position with one of the other guys. This was great as it allowed me to not only rest, but it was letting my ankle and leg calm down and feel better.

I had been there about 30 minutes when I got word to go back into the Base Camp. Our Senior Evaluator, the real 18C Engineer, had decided that since my leg was banged up, I would go back to being the Team Sergeant so that I could stay in the Base Camp and help plan the patrols and missions, several of which would be Engineer-oriented. While it did save wear and tear on my still healing leg, it also guaranteed that I would get very little in the way of rest for the remaining exercise, and that I would be very involved in all the aspects of the Command requirements of the student evaluations during the interaction with our Guerilla Force and the Auxiliary Forces.

It was not an easy exercise and I was physically and mentally tested to the highest degree several times during this iteration of Robin Sage. Looking back on it, while I would do it again, working through the pain and putting that type of pressure on my leg so soon after the injury occurred, really set my healing process back and it took almost a year before I felt like my leg was back to normal. It also permanently set my foot off kilter just enough to create some issues years later. Near the end of the field exercise, one of the evaluators was talking to me and I have never forgotten what he said. He said that after watching what I went through and just kept going, he doubted just how intelligent I might be, but he had no doubt that I had a heart of iron and one of the strongest wills to succeed that he had ever seen before. He shook my hand and congratulated me on not only being able to last the entire exercise, but to have done so well while serving almost continuously in a role as one of the Detachment's Leaders.

Soon it was completed and we were back at Camp Mackall. We went through the debrief and had our input on the required peer evaluation. I was looking forward to graduation and getting my orders to go to Group. I was hoping to get assigned to 5th Special Forces Group, which at the time was still located at Ft Bragg, NC. I had already had a couple of conversations with the Group's CSM Dennison, and he assured me that he had a slot for me on an SFODA as soon as I finished the Q-Course. While I was looking so forward to a new way of doing things, at least one of the members of my student Detachment was not. After all we had been through, one of the guys who was from a helicopter aviation background told the cadre that he wanted to quit and go back to his former unit, and stay in his former career field. The idea of doing what we had done in Phase 3 as a regular job was not what he decided he wanted to do from now on. He was separated from the rest of us and that was the last we ever saw of him. A Special Forces career was, and still is, not for everyone.

My First SF Mission, SFODA-591

After I graduated from the SFQC, I got my assignment. To my surprise, it was for 1st Special Forces Group way out at Ft Lewis, Washington. I had not been planning for that and was going to do whatever I could to get it changed. My family was already with me at Ft Bragg, NC. I had been stationed there before I even started my SF training, and I had planned on getting assigned to 5th Special Forces Group, which at the time was headquartered at Ft Bragg as well.

I immediately drove over to the 5th Group HQ and talked to the Command Sergeant Major of the Group, Joe Dennison. I had spoken to him before I ever started the SFQC, and I'd kept him posted on my status as I went through each of the Phases of training. When I showed him my orders for 1st Group, he assured me that he could fix it. He had a vacant position for me, line and paragraph number, and all I had to do was carry it back to the assignments section of the school house. The clerks working the assignments desk simply changed my orders and printed a new set out. That was that. I was now assigned to 5th Group and life was good.

Once I got to Group, I found out two things. First, I was assigned to SFODA-591 in Charlie Company, 3rd Battalion. Second, I found out that due to my still mending leg, I was being sent to work in a temporary duty position at the Group Engineer's office. The Group Engineer's office handled many things, some having to do with space utilization, future designs for office and training spaces, as well as managing a self-help store for fixing or building items in and around the Group area.

As the NCOIC of the Group's Self-Help section, I worked on things like security or safety installations or repairs, and even some required wood-working projects. The job also provided the Group's Command element with a "go to" individual for whatever could be done in-house. At times it was interesting enough, but for the most part, it felt quite unimportant and boring, especially for someone who just graduated and wanted to be on a Team. On the good side, it provided me with a great opportunity to build my leg back up, get some physical therapy through proper exercise, and get back to 100% rather quickly after doing a lot of damage during my too-early return to training in Phase 3. I was at the Group Engineer's office for about six-months, from June until December. I returned to my Detachment just in time to go into isolation and deploy on our annual Detachment evaluation. It was like learning to swim by jumping into the deep end of the pool. Literally.

SFODA-591 was a Scout Swim Team. This means the Detachment did almost everything a SCUBA Team did, minus being underwater while doing it. One of the primary means of infiltration was the over-the-horizon surface swim. This was when you entered the ocean far enough out that no one could see you and you could not see the coastline. Then, using a rope line that kept everyone together, you swam to a predetermined landing point on the shore, usually at night.

This was the infiltration technique that we would be using for our annual evaluation, or ARTEP (Army Training and Evaluation Program). The method we were going to use to get to our starting point, which could have been a ship, submarine, or even a parachute drop into the ocean, was by what is called a Helocast. This is when a helicopter flies above the water, usually at a 10-knot forward speed and 10-feet above the wave's crest, and the Team members exit the aircraft with their equipment by jumping out and falling into the water.

From that point the Team assembles, ropes everyone into a bud-line, and then the long swim begins by following a compass, among other techniques.

The complete concept of the operation was to infiltrate, move overland to a target, engage the target, exfiltrate to a predetermined point, catch another helicopter, and fly back to our base. We isolated at Ft Bragg and then flew to a point off the coast of Georgia, made the Helocast and maneuvered to the shore, where we moved overland into a military training area to interdict a target, and then moved via air back to Ft Bragg. The mission was supposed to be completed over a two to three day period.

Of course, a plan is always just the first step, and this well-planned mission hit a snag as we got ready to get wet. Our primary location to exit the aircraft was declared a no-go as we were about five-minutes away. There were two MH-60 Blackhawk helicopters from the 160th Special Operations Aviation Regiment carrying us in. We soon got the word, according to the safety team in the water (in boats) that the LZ was full of sharks. Real-world, not a training scenario. We had to fly to our alternate LZ, a few miles away. As we got ready to exit into our alternate LZ, we were told that the wave height was between 8 to 10 feet, which meant that we would be in rough seas. The fact that we had to go into our alternate because of sharks in the water was also playing in my mind. I wondered if there were sharks in the water at the alternate LZ also.

Soon we were told that we were two minutes out. Our rucksacks were setting on our laps, tied by a length of rope to our waste. We were wearing vests to keep us afloat, and had on dry suits as the Atlantic Ocean in December isn't very warm. We were wearing our other gear as usual, to include fins and masks for our swim. I was the 1st in the exit order from my bird. Having been absent from normal Team training the prior six-months because of my time at the Group Engineer's, I had never done a Helocast before. Before we launched, I'd been given a choice to meet the Team in the water and go in via the safety boat or join the Team in the Chopper and make my first Helocast count. Of course, I went with the Team. I figured gravity would bring me to the water and after that, it was all just swimming.

As I was told to get ready at the one-minute mark, I took off my headset and slid out the open door a little. The wind was pulling at my swim fins and forcing my legs to move in a circle. For a second or two I thought the wind would drag me out. As I waited for the Cast Master to tell me to go, I got the feeling that we were flying quite faster than 10-knots. Also, it was pitch black dark and I could neither see the horizon nor the ocean below. All I could see was my ruck on my lap in the dim red light of the aircraft's interior. Then the Cast Master yelled GO and pointed at me to exit.

I pushed my ruck off my lap and followed it out the door, past the wheel of the Blackhawk and into the darkness of the night. I felt myself turning upside-down and still, I had not hit the water. I continued to rotate forward and after what seemed like a long time, I hit the water feet first. Even with the buoyancy of my dry suit and vest, I went deep under the water and it seemed like a long time before I bobbed back to the surface.

When I did, I realized that the force of hitting the water had stripped some of my gear out of my pockets, but nothing of any value besides my 1st Aid kit had been lost. My ruck was still tied to my waste and floating, and I still had my M16/M203 rifle and grenade launcher. However, between the two aircraft, the Team was strung out much further apart than we had expected. Another thing I noticed, and was happy about, was that the ocean was calm. No waves at all. But what that meant, as I found out later, was that we had been told to exit the aircraft at about 20 feet above the water. Also, the birds were flying faster than they were supposed to be flying, again, about twice as fast as the recommended 10-knots.

It was quite the introduction to Helocasting and I never experienced one just like it again. Once we were all assembled and everything was accounted for, everything important anyway, we discovered that the Team Sergeant had landed on his rucksack when he hit the water and hurt his arm. We found out later that he cracked one of his forearm bones when he hit. Also, my dry suit zipper had ripped open about two inches during the crash splash and I now had December's ocean water inside my suit and it was a bit chilly. I figured it would warm up as soon as we started to swim. It did.

There were ten of us on the Detachment and we attached ourselves to the bud line two-abreast, creating a row, five pairs long. I was on the end, on the right side. The Team's Intel SGT was on my left. After swimming for a while, I noticed that we weren't seeming to make much headway.

Apparently, our alternate location for the LZ placed us on the southern side of an island that was about a mile or two off the coast. The current on the southern side was flowing in a different direction than the currents flowed on the northern side of the island. It seemed that we had accounted for the tidal current, but not the way the ocean reacted as it moved around the island. We kept swimming, but we were not making very good time.

It so happened that in the area where we were swimming, there was a florescent alga in the water. As we moved through the water, we were leaving a long trail of blue light. I joked to my swim bud that it reminded me of a sign pointing to us that was like an advertisement for a diner. About that time, we actually saw a large fin cutting through the water near our position, and then it went beneath the surface again.

Given that this was a training mission, we still had a safety boat following us a bit behind and off to one side. My swim-buddy, the Intel SGT yelled to the boat that there may be a shark swimming near us and to consider picking us out of the water. We were told that the fin belonged to a dolphin (more than likely) and not to worry about it. Easier said than done. We continued to swim.

After a couple of hours, we almost got run over by a fishing boat as it came speeding past us with no running lights burning at all. Right behind the 1st boat came a second, and it had plenty of lights because it was a game warden or some version of water police. Seeing our safety boat in the light, the 2nd boat stopped and the captain asked what in the world it was doing out in the water (with no lights), and then he saw us swimming. His story was that he was giving chase to someone doing something wrong, for us to be more careful, and then he was gone, chasing after the other boat again. This night was not getting off to a good start.

About that time, our Team Leader decided that we had to make a change in plans as it was starting to become obvious that we would not reach our rendezvous point in time to make a link up with the auxiliary for our ride into the target area.

The safety boats were contacted and after calling the rear for further instructions, it was decided that we would get a ride from the boats to our initial coastal infil point. However, the boat operators got lost and couldn't find the location due to the tide coming in and covering some of the markers or landmarks they had looked at earlier during their preparations for the exercise (during low tide).

At that point, the Detachment was taken to the island off shore that had been causing all the problems with the currents, and we were dropped off and told to establish a perimeter until morning. At that point we were able to repack our equipment and get rid of the water gear as we pulled maintenance on our weapons, ate a little chow, and got a little rest.

As the sun started to come up, we loaded into the boats and headed back towards the shoreline and stopped at a small dock where some fishing boats and a small store were located. As we unloaded the boats on the dock, our radio operator dropped one of his radios out of his rucksack pocket when he lifted it up to the dock, and it hit his head, cutting a gash as it did so. We went into the small store and talked to the owner as the medic stitched up the cut on our Teammate's head. The storekeeper asked if we had been out in the water overnight and then went on to say that he would not have been in the water for anything. According to him, the fishermen had pulled a huge Hammerhead Shark out of the bay only the day before and he figured it had to have a mate out there somewhere. Of course, me and the Intel SGT were pretty sure we had seen exactly that the night before when we saw the fin moving through the water right before it submerged and disappeared.

Soon we were in the back of a U-Haul truck and headed towards the military reservation's boundary where we would disembark and continue by foot to the target. The Team Leader was riding up front to make sure we went where we were supposed to be going, but the rest of us got some much-needed sleep as we rode in the darkness of the truck box for the next hour or two.

Once we got stopped and out of the truck, we started moving through the edges of the swamp and things began to feel normal again. I was used to patrolling on the land, even wet, muddy, swampy land.

As luck would have it, the target was a bridge that was guarded by a squad of infantrymen. Since I had planned a bridge destruction mission during my training in the SFQC, I was all over this mission with plenty of fresh school-house knowledge. My rucksack had the charges that I had already built during isolation using the target folder, so all I had to do was confirm the target was the same as the information in the target folder had indicated. I had plenty of extra, just in case.

We soon discovered that the bridge was in a very dense and very swampy area with only one or two obvious lines of approach. We decided that we would not be taking either of the obvious lanes. The infantry squad had been there a couple of days and they were beginning to get bored and lax in their security. It cost them dearly. We came in through some of the most unforgiving swamp I have ever crawled through, but by the time they knew we were there, the evaluators had called them dead or wounded beyond any capability to defend their position. I set the charges and we ran out of the area and back into the thickness of the swamp for additional security.

Once we stopped, most of us were sopping wet again. It was getting cold and we didn't want to start a fire so close to exfiltration time, so we got some of our dry or drier clothes out of our rucks and continued moving towards the Pickup Zone. At this point we had been going very hard for a few days and everyone was tired. It was a hard movement to get to where we were supposed to be picked up by the 160th SOAR and by the end of the movement, it had turned into a real challenge to maintain security and keep a tactical attitude going. Finally, we were on site and as we neared the 4-minute window for the pickup, the 1st bird's group moved into position to board upon landing. I was going to be riding in the 2nd bird, so I was out on perimeter with the others, pulling security for the 1st group. At one minute into the window, we could hear the whup-whup of a bird inbound. It settled down as a big ring of static electricity formed a circle of light in the night at the ends of the spinning blades over the top of the Blackhawk and the 1st group piled into the bird. Oddly, it stayed on the ground and then someone flashed a light in our direction. It meant we were to also load up. Quickly we ran to the bird and jumped in on top of the others as there was little space left open with the bladders of extra fuel in the passenger compartment. All accounted for, the bird lifted off and headed away from the exfil point. The crew chief, as usual for the 160th, handed us a box of doughnuts as soon as we got to our flight altitude and started flying towards the coast. Apparently, the 2nd bird broke down on their way to pick us up and had to go elsewhere for maintenance.

Soon we were flying over the ocean and we turned towards the north to head back to Ft Bragg again. It was dark, so the lights on the water got my attention as soon as we started getting close. It was an oil platform off the coast and the lights were reflected in the water. It was also being used as a link-up point for the helicopter in which we were riding, and the 2nd bird that had detoured elsewhere for maintenance. The pilots were flying with NODs or Night Vision Devises, and apparently didn't see the other aircraft flying toward them from about our 8-oclock position. I watched it fly towards us and neither bird seemed to be moving out of the way of each other. I grabbed the crew chief by the arm and pointed at the other helicopter. He said something very excitedly into his commo and the pilot jerked the aircraft upwards as the 2nd one flew beneath. They never said, but I guessed they had not seen each other and we had come too close for comfort to running into one another.

We continued back to Ft Bragg and soon we were back in the isolation facility for our post-mission debrief. There was chow and coffee waiting for us in our isolation room, along with a dry, and clean, uniform. The mission had been a great success as far as everyone was concerned, even though it started out as rough as it had. I made a good first impression and was glad I didn't have to wait very long to be able to prove myself to the Team.

As it turned out, this was my first and last mission with SFODA-591. While attached to the Group Engineer's office, I had managed to get approved for a HALO or Military Free Fall School (MFFS) physical and passed, in spite of having a leg still mending. You had to have a current physical before you could be selected to attend a specialty school like MFFS or SCUBA school. About that time, the Company got a school slot for MFFS and I was the only one with a current physical who was also available to attend the school. I passed the course and was assigned to SFODA-594, the MFF Team in the company as soon as I returned from the course.
Not too long afterward, our entire Company was deployed into a few southwestern US areas in support of various training events and some Joint Task Force counter-narcotic missions.

During one of these missions, my 1st Team, SFODA-591 was involved in a helicopter crash and several of the members, to include the man who took my spot as the Team's junior Engineer, were killed.

I attended his funeral as part of the Honor Guard and always thought how lucky I had been to have gotten assigned to the MFF Detachment after going to MFFS. This was one instance where passing a course ended up saving my life. It was bad for the other guy and his Team, but no one knows what really impacts the events that happen all around us. I and members of my new Team had been on that same helicopter several times during the same deployment before it crashed. It could have been us just as easily. This all helped me understand to this day, that not only is life good, it shouldn't be filled with wasted time.

Jordan, Ebony Wood 1988, SFODA-594 (MFF)

A Biblical land, Jordan; officially the Hashemite Kingdom of Jordan. Located east of Israel along the Jordan River and against the Dead Sea to the west, and touching the Red Sea to the extreme southwest, the land of Jordan has been inhabited by mankind and several kingdoms since the beginning of humanity on earth.

For my first deployment on my new team, SFODA-594, the Company's Military Free Fall Detachment, I deployed to Jordan with the 5th Special Forces Group in 1988. Our mission was to train and participate in field exercises with the Jordanian military from their headquarters near Amman, the capital of Jordan. Our Foreign Internal Defense (FID) mission was primarily to conduct instruction and joint training exercises to include infiltration techniques and light infantry tactics and techniques to elements of the Jordanian Special Forces.

This was my first deployment with Special Forces to a foreign country. After some initial training we presented to the host nation soldiers, we started planning a large operation that would take us from our location near Amman in the northern area, to the vicinity of Al Jafr, about 200 miles south and a wide-open desert area with mountainous terrain.

The mission was to be a hostage rescue operation and included a parachute infiltration, a long tactical foot movement of a Battalion sized element, an extraction from the mission site using host-nation aerial assets, and exfiltration from the rendezvous point utilizing the USAF Special Operations Combat Talon MC-130, via the Blackbirds support squadron.

As the mission started, I and a portion of my Team, SFODA-594, boarded the designated MC-130 with the members of the Jordanian Special Forces we were leading on the operation. The static-line parachute insertion was scheduled for the evening, once it became dark. As we were flying into the area and had started to receive jump commands from the Jumpmasters, we started to get some turbulence from a storm front that had started to move in.

Both the left and right side of the aircraft had jumpers standing up and we were about to hook our static lines to the anchor line cables. Almost everyone had rucksacks and weapons. I was at the end of the stick, or line of jumpers on the right-side of the aircraft. I was going to be the last one to exit from my side once we got the Green Light for the jump. I could see that the turbulence and rough flight was getting to some of the Jordanian jumpers and between the flight and the darkness outside, the level of motivation was dropping rapidly.

The Jumpmasters gave the hook up command and everyone complied as the static line hooks were snapped onto the anchor line cable overhead. Then as the Check Static Lines command was given, the hand signal used caused some problems. The US jump command hand signal for Check Static Lines is the OK signal with your thumb and forefinger forming a circle with each hand, while moving your arms up and down and pointing towards the jumpers. This hand signal caused an issue since in the Arabic culture it was also a gesture of the evil eye, and was seen as a sign of placing bad luck on someone.

As soon as the signal was given by the Jumpmasters, many of the jumpers began complaining and several unhooked from the anchor line cable and sat back down. Once a few had sat down, several more sat down. By now the ride had started to get much rougher and the turbulence was making standing up more of a challenge. At this point, we received word from the Drop Zone Safety Officer that the jump was being cancelled due to high winds. This ended up being a good thing and prevented a real mess inside the aircraft.

After we returned to the airfield and loaded into trucks, by the time we got on the ground in the area of the original DZ, it was about midnight. We still had a long walk ahead and it was easy to see that the movement was going to require extra effort on the part of the Jordanians.

After a couple of hours into the movement, it was apparent that the Jordanian officer leading the unit didn't know exactly where he was going. This turned out to be an indicator of a common shortcoming. Also, as I was in the rear, I noticed that some of the soldiers were dropping heavy items to include canteens and ammo in order to lighten their load. Soon after, one of the US Team members took charge of the patrol and we began navigating towards our planned ORP, or Objective Rally Point and established a security halt as we reconnoitered the ORP and sent out a security patrol.

Once the morning had broken, the mission continued and the raid was conducted resulting in the successful rescue of the hostage. The main element began its movement back out of the area towards the location of the original DZ in order to get picked up by the trucks that had dropped us all off the night before.

The smaller party comprised of the US SF and a few Jordanian SF, along with the hostage, were almost immediately picked up by a pair of old Jordanian helicopters, French-made Super Pumas, at our raid location. The choppers flew us to a major highway about 30-minutes from our pick-up point.

The highway was the primary north-south transportation route that ran through the mountains. We had identified a straight section about a half-mile long as our exfiltration point. As we flew into the area aboard the Pumas, a Blackbird crewed USAF Special Operations MC-130 was landing on the highway. The Jordanians had blocked the road for a few minutes to allow this to take place. The Pumas landed directly behind the MC-130 as it was still taxiing to get ready for a quick exit. We ran to the tailgate, which was lowered to about a couple of feet off the pavement, and I counted each of the Team members as they jumped into the plane. It had started rolling as I jumped onto the tailgate myself, as the last patrol member to board. I moved forward to the floor of the aircraft and grabbed onto a strap as the bird lifted off and started to raise the ramp. It was smooth sailing after that and we returned to base for a debrief and hot wash prior to generating the After Action Report.

A Patrol Along the Israeli Border

During our mission in Jordan, we accompanied the Jordanians on several actual security patrols. My team was divided between several areas and I, along with the Senior Medic on my team, went along with a Jordanian patrol that was going to be conducted near the border of Israel. Our role was mainly to observe and provide feedback as appropriate, primarily during the hot wash upon completion and return to base.

We departed our base in the back of a couple of the local unit's trucks with a Jordanian SF Captain and about 24 soldiers. One of the first things I noticed was that their rucksacks were much smaller and lighter than the rucks my partner and I were carrying. This was to become even more noticeable later during the patrol. The launch point was in the mountains and the roads the trucks were driving up were practically smaller than the trucks themselves.

I moved from my seat in the back of the covered truck to the tailgate and actually sat there watching the rear wheels as they came within inches of the side of the mountain road. My plan was to bail out if I felt the truck start to go over the edge of the cliff. The Jordanians didn't seem to mind being in the truck on these roads and I was reminded of the old saying, the worst ride is better than the best walk. Still, I kept my eye on the cliff, or what I could see of it since it was dark.

After an hour or so, we unloaded in what seemed like the middle of nowhere and started getting ready to begin the patrol. As we spoke to the Patrol Leader and discussed our route, I noticed that he wasn't carrying any water on his gear, nor any ruck sack. I asked him if he had water and he stated that his water was carried by one of the soldiers. Once more, I was reminded that we were not dealing with soldiers who thought like Americans. I also started to look for antennas and saw none besides the ones that I and my partner carried for our radio equipment. I asked and found out that they didn't have any radios on the patrol.

Soon we got started and my position was towards the rear of the patrol. Within minutes, we were on a goat path that required everyone to be in a line, ducks in a row style. On one side, the mountain went almost straight up in the air and on the other, it quickly dropped off into the darkness with only the top of what was either clouds or fog showing from below. I had no idea how far the drop off went to the bottom and I wanted to be very careful to not find out.

At the first security halt, we moved towards the Patrol Leader's position and talked about some of what we had seen and what to expect as we moved forward. Security was almost nonexistent and when asked about the route ahead, the CPT seemed unsure. I asked to see his map and he had to find it in his ruck sack. It was obvious that he had just been following the paths as we went along. Now, he was unable to locate our position on the map and simply stated that we were headed towards the next village and would see the lights as we got near.
While looking at his map, I noticed that Israel was not identified by name. The border was noted, but it was shown only as a territory with a disputed border. I also had my US map and had been following our route with it. We knew where we were, but it was obvious the CPT only knew about where we were. This continued for a few hours and finally at the next security halt, the Senior Medic told the CPT that he was lost and that we needed to change our planned movement and get back on course. At that point we had passed the trail to get to the town where we were to link up with the trucks again and had to lead the patrol the remainder of the way.

This was when the extra weight in our ruck sacks really started to be noticed, but even then, we were doing much better than most of the Jordanians. It was pretty clear that they must not have been used to such a long patrol and they were quickly wearing out. Thankfully, for them anyway, we were able to quickly get back on the right path and soon we saw the lights of the village in the distance. Once we got back to the trucks and returned to the base, our chow of goat bones and rice was tasting pretty good. It felt good to be back at our base camp, as humble as it was.

Speaking of Our Base Camp

I need to mention our base camp. As mentioned earlier, Jordan is a Biblical area and it has been occupied and travelled over by many different groups, for many different reasons, for many eons. Our base camp for this operation had been established in what was the ruins of an old Crusader fortress. Not far off a major road but still almost unknown to most of the people driving by, this fortress had seen better days. There was a main structure that consisted of a few rooms and it had obviously been used as a place to temporarily stable goats or sheep for many, many years. Probably centuries. Between the rooms, all created from stones that had been set in place during the Crusades, we managed quite well to stay out of the wind and even a little rain and snow.

One unique fact about this set of ruins was that there had been a well dug from inside the fortress, down into the mountain when the fortress was being used as a military stronghold. In one of the rooms inside the portion still standing was a set of steps carved into the rock and the steps descended down a deep hole that had been dug which was over 100 feet or more. This deep, step-lined shaft ended at a small basin of fairly clear, spring water. This small carved basin at the bottom of this shaft was still staying full of water that allowed the knights of old to survive. It was quite a testament to the ingenuity and capabilities of the ancient people who built it in order to maintain a supply of water in the middle of the desert, year-round. Of course, this was in the same region where according to the Bible, Moses struck a rock with his staff and opened a spring of fresh water.

And while on the subject of bases, or places we stopped overnight, during another patrol, we stopped and set up a small camp in an olive tree grove that had to have been a few thousand years old. The locals told us that this had been an active olive grove during the time of Christ. I spent several hours that night just lying there under my poncho liner looking at the rows of olive trees and up into the heavens, wondering who else had stayed in this spot over the years and what, if they could only talk, would the ancient garden inhabitants be able to enlighten me with, regarding history and perhaps, even tales of the Son of Man.

We conducted several more operations, to include an airborne operation where the Jordanian soldiers actually jumped from the aircraft allowing me to earn my Jordanian parachute wings. As we neared the end of our deployment, we got a chance to stand down for a weekend and go do the tourist thing near a couple of interesting places, one called Aqaba, and one called Petra.

Touring Real History in Jordan

Petra is a very ancient city, or what remains of a city, that got its start around 7000 years BC, perhaps earlier. What we saw during our visit to the site were almost unbelievable relics of ancient history. To enter the place, you had to walk along a narrow channel through vertical cliffs for about a half mile. Once you exited the narrow passageway, the valley opened and in front of you stood a scene that millions had viewed during an Indiana Jones movie. This temple, or treasury as it's also called, is carved right into the sandstone mountains lining the valley along with many other structures and ancient homes. This entire place is also known as the Rose or Red City because of the color of the sandstone. However, like many real-world facts, they failed to meet the expectations created by fantasy stories concerning the same subjects.

For instance, the treasury, which seemed to be an entrance to a wild underground system of caves or tunnels in the various movies, actually held a simple and quite small rectangle room with a couple of small storage nooks carved into the walls. Still, the fact that this had all been accomplished by people so long ago and survived intact was amazing.

Further into the valley there was an open-air stadium of sorts, with what had been a stage for plays or perhaps public meetings. It was carved from the mountain as all the other structures and you could hear the words being spoken on stage from any of the benches carved into the rocks surrounding it.
It's a place that has become much more popular than when I was there in 1988. For centuries, it was forgotten to all except the local people who tended their livestock and lived their Bedouin lifestyles throughout the region.

After departing what has to be called an ancient wonder of the world, we traveled to Aqaba. This is another city that has, not only a long record of existence, but is an important part of world history.

It is said that Mohammad of the Islamic religion led his forces into this area in what's called the Expedition to Tabouk in 630 BC and received a high level of payment in exchange for not destroying the city.

Aqaba was the focal point for trade between the Middle East and Africa for thousands of years and continues to be an important port allowing Jordan to have access to the rest of the world through cargo ships traversing the Red Sea and beyond.

Given the short amount of time, along with a quite-restricted movement ability, I was unable to see much of the city. However, I did stop at a couple of coffee shops and enjoyed the local espresso style of strong, thick coffee, along with some very unique pastries that were something like deep-fried doughnut holes filled with dates and nuts.

For me the highlight was being inside the ancient walls of the city in Petra and walking through some of the small carved rooms surrounding the larger monuments and structures. I enjoyed the visit so much because it allowed me time to consider what all had happened in these rectangle-shaped spaces, carved from the sandstone, so many centuries ago. Back when they were rooms of some unknown family's home. Real history.

Bad Juju and a Helicopter Crash

Sometimes, you see something that sets your teeth on edge. As my Detachment boarded an old Jolly Green Giant, I experienced one of these times. The HH-3E helicopter landed on the drop zone in Arizona. We were going to be jumping from it for some Stand Off missions, flying under canopy between 10 to 15 kilometers before landing. We had jumped from helicopters before, a lot actually. But this time, it was different. Pasted on the side of the aircraft next to the cargo area door, was a bumper sticker. A blue sticker with white block letters. It said simply, "We'll get along just fine as soon as you realize I'm God." I remember looking at it as we climbed aboard and telling the Crew Chief that the sticker was bad juju.

We flew that bird for two missions and everything went well. I soon forget about it and we continued to train as scheduled during our deployment.

A few days later we got word from our HQ that the operation was cancelled and we were to return to Ft Bragg. That's about the time we found out that the same HH-3E Jolly Green Giant had been transporting one of the other Teams across the desert when it had crashed and killed everyone aboard.

One of my best friends was the medic on the Team and he had been on the helicopter and then got off just minutes before it flew away in order to drive the ambulance to the new training location. He was suffering some pretty intense survivor's guilt. The Detachment had been split between two choppers and the one in trail, the one following the lead bird, is the one that crashed.

After the lead chopper noticed that it was alone, it turned to search for the other helicopter. They found it where it had crashed and burned in the desert. The rucksacks of the Team members were scattered around the crash site in a circle where they had been slung out through the ramp and side exits. The Special Forces community had been given a recent directive that personnel riding in a helicopter had to be strapped in during flight operations if they were not participating in an Airborne Operation. Apparently, the aircraft had experienced a mechanical failure that caused it to spin with the main rotor as it fell to the ground. The rucksacks being loose, flew outside and made a ring around the site. It was quite eerie, as the rucks were in perfect shape, but everything else was gone, buried in the sand beneath a melted airframe and engine parts.

The SFODA members were from the Detachment that I had been assigned to before I was transferred to the MFF Team. The Engineer that had been assigned to my old position was one of the soldiers killed in the crash.

Our Company reassembled and returned to Ft Bragg. We spent the next several days going around the country as Honor Guards and members of burial details for our Teammates who died in the crash.

I always think of the bumper sticker being on the helicopter whenever I think of the crash. I know the sticker didn't have anything to do with the crash itself, but I also felt that you shouldn't tempt fate either. Sometimes a joke can be more than simple bad taste, it can be bad juju.

La'Afrique, Land of Adventure

During my tours in the 5th and 3rd Special Forces Groups, I spent a lot of time in Africa. Our primary areas of operation were spread throughout Africa, as well as much of the Middle East. The various Battalions to which I was assigned however, focused mainly on the parts of Africa that were south of the equator, what we referred to as the sub-Saharan regions. My specific Companies had missions within the equatorial and southern regions.

My multiple trips into our mission areas brought me to the countries of the DRC or Democratic Republic of Congo (formerly Zaire), Zimbabwe, and Botswana. During my missions into these countries, we at times were interested in what was happening just across the border into some of the other nations, such as Angola, Zambia, South Africa, Namibia, and Mozambique.

Our primary reason for traveling into these various areas of Africa was known as FID or Foreign Internal Defense missions. Foreign Internal Defense is defined by the Army as "participation by civilian and military agencies of a government in any of the action programs taken by another government or other designated organization to free and protect its society from subversion, lawlessness, insurgency, terrorism, and other threats to its security."

This was another way of saying that when we went in, we would work with the host nation's military to better enable it to protect its government from an enemy. Besides UW or Unconventional Warfare, at that time, FID was the backbone or premier mission of Special Forces. Later, as SF began fighting in Iraq and Afghanistan after the attacks of 9.11.2001, the Counter Insurgency or COIN mission would get the attention of the world, but COIN was just one sub-mission of the FID tool set.

Of course, missions usually have names, some more cool than others, and I was involved with quite a few missions in the African Continent. Names like, Flintlock, Bondeko, Mabote,

Lancer Jaguar, and Falcon Cherokee. Some sounded African, some didn't. Some were missions that included other areas besides Africa, and some were limited to only our Africa-based areas.

Zaire (DRC) Kinshasa 1988 Bondeko and Kamina 1989 Mabote SFODA-594 (MFF)

Zaire was a country that simmered just below the surface and could erupt easily. Politics and tribalism, mixed with a history of colonization and foreign interference, set the stage in Zaire to make it a perfect place for US Special Forces to come and work their FID, or Foreign Internal Defense mission magic. Zaire was also the birthplace to my SF nickname of Congo. First just a short look at what made Zaire's history and future so unique.

The capital of Zaire was Kinshasa. Kinshasa sets along the southern side of the Congo River, directly across from Brazzaville, the capital of the neighboring country of the Republic of Congo. Kinshasa, the first place in Zaire to which I was deployed, was initially named Leopoldville, after King Leopold II from Belgium. In the late 1800s, the Belgians controlled the territory that would become Zaire and called it the Congo Free State. Once the Belgians colonized the country, it was known as the Belgian Congo. Later it became the Republic of the Congo and the newly elected prime minister began to become influenced by the former Soviet Union.

The USA and Belgium wanted to prevent the USSR from gaining access to the country and supported a rival to the prime minister, resulting in his 1965 assassination and replacement by Joseph Desire Mobutu. He was the Belgian and US supported political leader and was still in charge when I deployed there in 1988. That stated, the USSR and China, as well as other anti-US and anti-NATO nations provided equipment and other assistance until the end of the Cold War, after which Mobutu allowed the military's readiness condition to degrade in order to help him stay in power as he feared a military coup.

Mobutu was a traditionalist and changed the names of most places from European names to African names. Leopoldville became Kinshasa, while the country's name was changed from the Republic of Congo to Zaire. Mobutu turned out to be a corrupt prime minister and did a lot of bad things, but managed to maintain power because the country flourished in many ways, and the city of Kinshasa became a mega-city, the 3rd largest in Africa.

During the late 1990s, Mobutu's regime was overthrown by a rebel uprising comprised mainly of his former soldiers, many undoubtedly, of whom had received training from my Team during our deployments. The French also maintained a permanent military presence in Zaire and were often at odds with our mission requirements during our visits during the 1980s.

After the 1997 downfall of Mobutu, Joseph Kabila was elected president and renamed the country Democratic Republic of the Congo or DRC as it is known today. For now.

The military personnel we trained were known as the FAZ or Forces Armees Zairoises. We simply called their soldiers Zairoises, pronounced Zy-waz. We worked primarily with the 2nd Groupement in Kinshasa, which was comprised of two Brigades of Parachute Infantry and one Brigade of Light Infantry known also as Commandos. We also worked with their elite unit, the Special Presidential Brigade which was several Battalion sized Infantry units. All of these units were located near Kinshasa and Kamina.

As you may have noticed, there is a lot of references to the name Congo. One of my duties during our FID missions was to conduct Physical Training with the Zairoises. I would normally help call cadence during our morning runs and I also could replicate a very close to original, Tarzan Yell as we would run around the compound. Eventually both my Teammates and some of our counter-parts began to tag me as Congo Bill, and it stuck. It quickly became simply Congo and I was known throughout several SF communities as Congo for long after I retired.

The Magic Rock, Kamina, Zaire

Kamina is home to an old Belgian air field that was our base of operations during our Mabote exercise. Located in the southern region of the country, it has seen many uses and is generally much more rural than any of the locations around Kinshasa. While I was deployed to Kamina, there were several large hangers that were reported to be controlled by the CIA. These hangers were also rumored to be used during operations when the CIA would fly weapons into the Kamina Airfield, to be flown on further, to an airfield in Angola, just to the west. The weapons would be used by UNITA, a guerilla organization fighting against the Marxist Angolan government, which was being backed by the Cubans, among other US adversaries at the time.

While training with the local Zairoises around Kamina, we touched on many light infantry subjects, to include movement techniques and formations, many of which we would conduct during our rehearsals or demonstrations. During one such event we had one of the more sociable personalities, a Zairoises Company Commander with us. He was a very experienced military officer and besides training in Zaire, he had also attended the Royal Military Academy of Sandhurst. This is like England's version of the US Army's West Point Military Academy.

Knowing some background and having talked quite extensively with him, added to my surprise when I heard him speak of a magic rock.

As my Detachment was showing him some of our immediate action drills and other patrolling techniques, we were maneuvering through the flat grassy areas between the inactive taxiways of the airbase. The grass was about knee to thigh high, and the ground was flat and otherwise, obstacle free. It was a good place to move around and a lot easier place to take a knee during our stops than the concrete would have been. After having demonstrated a few of our formations, we gathered back on the concrete taxiway to discuss what we had been showing to him and the other soldiers.

At some point in our discussion we were told that we should conduct any more maneuvers on the wide taxiway rather than out in the grass. The answer to our question of "why?" was a bit surprising. It was because of the local vipers that apparently were quite numerous in the area. One of the local snakes is called a Gaboon Viper and it apparently liked to hunt for rats in the grassy fields between airstrips. While this snake is not an overly aggressive snake, it is very deadly if it would bite. We were told that it was as bad as a King Cobra and if you were bitten, you were guaranteed to die quickly if it injected you with venom. Of course, we decided that after our warning, we would no longer use the grassy areas as a training area and would stay on the concrete as the locals did while they practiced our techniques.

Now, while this was all a bit surprising, it was nothing really out of the ordinary. As I've already mentioned, this Zairoises officer was a highly educated person, one whom would be expected to adhere to what we would consider normal "western" ideas. When asked about how they treated anyone who might receive a bite from one of these poisonous vipers, he told us about the magic rock. We thought he was joking, but quickly discovered he was quite serious.

He, and we assumed most if not all the other Zairoises soldiers, carried a small flat rock in a pocket at all times, at least while they were in the bush. This rock had been blessed by one of the local medicine or witch doctor types of people found in the nearby villages. If bitten, the victim would place the rock over the bite and it would magically prevent the venom from causing death. Which would be a good thing because there was no anti-venom available for the average villager or soldier who might get bit.

Since the viper was not real aggressive, it would sometimes bite when stepped on, but not inject venom. If someone got a "dry bite", the magic rock would work. Or, as the Sandhurst educated Zairoises officer told us, it was because the rock was really magic. You can decide. And no, we didn't start carrying magic rocks, but we were quite careful to keep an eye out for the snakes.

Soon after this, one of the local soldiers killed a Gaboon Viper inside our compound. They used it as part of their dinner meal that evening and had invited a couple of our guys to eat with them, as they usually did. One of them ended up with food poison, not from the snake's venom, but just from the general lack of hygiene in the kitchen. At least that's what our medic told him.

Magic Lights, Clowns, and Enemy, Oh My

Airborne operations inside Zaire was a normal part of our own training, as well as joint exercises with our Zairoises counterparts. A couple of the jumps made near the city of Kinshasa stand out, not as much for the jump, but because of some of what occurred around the periphery of the operations.

One of our night MFF or HALO jumps, we landed on a DZ a little south of the Kinshasa International Airport. This was still a rural area, but there were a lot of locals who worked and lived within and around a military compound in the vicinity of the DZ. We made the jump with all of our combat equipment, as was the normal procedure for our SFODA, especially during jumps in foreign countries. For various reasons, when jumping at night we would mark our equipment in several places by fastening Chemlites, or glow-sticks onto things like our helmet, altimeter, rucksack, and sometimes our leg. We used these lights as not only identifiers to see each other in the air, but also for after we reached the ground, at least initially during our administrative assembly and equipment accountability procedures.

We had made a late jump, and I was surprised when I hit the ground to find several kids on the DZ waiting for us to land. The local village children had come out in force to help us. The Chemlites, however, quickly became the center of attention. Apparently, they had never seen such a thing before and the Chemlites quickly became known as magic lights. I was glad to give all of my lights to the kids and they ran off into the darkness as happy as could be with their new toys.

A couple of days into the mission after we'd made the night jump, we had a visitor come to the training site. It was one of the kids from the DZ who had gotten one of the Chemlites. It was obvious that he had taken it home and the next day and built a wooden box in which to mount the light. It looked very much like a lantern. Only by now, the Chemlite has stopped glowing. He wanted some batteries for the light. He was sent away with the hope that he might be able to gather more lights should we conduct any further night parachute jumps during our visit.

During another parachute operation, we were jumping with a locally rented DC-3 aircraft and pilots. The pilots were expatriate Belgians, left over from the old days of Zaire being colonized. We had loaded a lift of Zairoises static-line parachutists into the aircraft. There were no seats in the plane, so everyone was setting on the floor wearing their parachutes with a length of rope tied across their laps as seat belts to keep them from sliding around during take-off.

Before the pilots started the aircraft, they turned towards the troop compartment from the cockpit to give us the safety briefing. As they turned around and faced us and the Zairoises, we could see that they were wearing crazy clown masks. This immediately unnerved about half the passengers in the plane who were already nervous about making a parachute jump, something they didn't do very often. The Americans thought it was quite hilarious, but we did have to work a bit to calm our jumpers down and get them to pay attention to the pilots and focus on the jump we were about to go make.

Later on, during another mission, we were making a high altitude jump about 200 miles southwest of Kinshasa, in the vicinity of a city on the Congo River named Matadi. This area was also very close to the border with Angola. In the late 1980s, Angola was a central player with its allies being Cuba and the Soviet Union, in the ongoing Cold War. With thousands of Cuban soldiers in Angola, our very presence in Zaire, especially being that close to the border was likely seen as a provocative act.

We enjoyed the fact that while in free fall, we could look into Angola proper. Although we never crossed the border, or even saw any Cuban soldiers, it nevertheless remained an interesting period of time and added a layer of realism to our patrols and other training events we shared with our Zairoises troops during the FID deployments.

Gaborone, Botswana, an I-TOW MTT

One of the more interesting missions was a FID mission during 1989 into Botswana that was directed (and funded) by an organization called SATMO or the Security Assistance Training Management Organization. SATMO evolved from the Special Operations Community and eventually became the US Army entity or USASATMO that it is today.

I was to be the Mobile Training Team or MTT's Team Sergeant. I would also be the primary instructor for the three iterations of training that we would conduct upon deployment into the theater. The team consisted of four personnel from my Company, a Warrant Officer as the Commander, myself as the Team Sergeant, and two other SF NCOs. I was a senior Staff Sergeant (E6) and I had a SSG Medic and a SGT (E5) Engineer.

Our mission was to instruct and train elements of the Botswana Defense Force (BDF) assigned to the Infantry Brigade at Francistown on the I-TOW (anti-tank) system. The Improved Tube-launched, Optically-tracked, Wire-guided (I-TOW) missile system would be mounted to a Cadillac Gage Commando V-150 wheeled armored personnel carrier. The I-TOW weapon system had not been used with this particular vehicle before, and the BDF was getting both the weapons and the vehicles at the same time. My Team would be providing the initial training which ranged from individual operation and maintenance requirements, to strategic tactics for the unit.

As an additional requirement, and mainly handled by the Team's Commander, we were required to interface with personnel from the US Embassy in Gaborone, and various Host Nation VIPs, to include the Commander-in-Chief of the BDF. This mission was made doubly important because it was an initial entry mission for US Forces into the country of Botswana. The BDF was not a remnant of a previous colonized force, such as the British or French, as were so many other African military forces. The BDF was a product of an originally created African force of paramilitary police, formed from the Botswana Mounted Police, a sub-unit of the British South African Police prior to the BDF being raised in 1977. When my

Team deployed to Botswana, the former Soviet Union and China, among other military organizations, were represented in Botswana and occasionally would show up unexpectedly at our training sites. This would cause us to stop training and notify the chain of command.

Prior to conducting the mission, SATMO provided several months of training for me and the other members of the Team. We attended training on the I-TOW missile system itself, at the Redstone Arsenal in Huntsville, AL; the I-TOW Platoon Leader's Course for tactical and operational level training at Ft Benning, GA; and a familiarization opportunity for the Cadillac Gage Commando V-150 in Detroit, MI.

Once we had received the training, we began to develop the instruction curriculums, which included creating the operational instructions for the weapon system and the new vehicle carrier.

During the time that we deployed to Botswana, there was an ongoing and serious issue between the BDF and the SADF, the South African Defence Force. South Africa was still under the apartheid form of government. The SADF was in operation until 1994, after South Africa abandoned apartheid with the election of Nelson Mandela.

Prior to my Team's arrival into Botswana in 1989, the majority of white men wearing camouflaged military uniforms were generally members of the SADF. Uninvited, these soldiers would cross the border into Botswana and attack elements of the ANC or African National Congress that had taken refuge in the country. This led to one of the more interesting occurrences to happen during our training.

We stayed at a hotel in the capital city of Gaborone during the nights and weekends and drove the several miles into Francistown to conduct the training each morning. During our trip we would usually stop at a small gas station that also sold local fresh foods and baked goods. After a few days, as we were getting ready to leave the station and start again towards the military base, I noticed a large group of men walking towards our vehicle. They had almost surrounded the parking area and some of them were carrying clubs.

As they began to walk closer, and we were back at the vehicle about to quickly drive away, I yelled out to the man who seemed to be leading the group our way. I asked him what he wanted. He looked at me a little funny, and then stated in a questioning way, "You are an American?" Apparently, my southern accent had given me away. I told him that yes, we were all American soldiers and we were in Botswana working with the BDF.

Once he heard this his entire demeanor changed and he relayed the information to his comrades, upon which they all started smiling and giving us many "good morning sirs" and started walking with much less hesitation towards where we were parked. We were suddenly local celebrities.

As we talked with the folks who gathered around, we discovered that they had seen us come in and had assumed that we were members of the SADF in the area looking for some local ANC supporters, and they had come to chase us away. Or worse. Now that we were known as the Americans, we were greeted and treated much more nicely than what had been quite nice already, prior to the crowd showing up.

In another example of how far from home we actually were, we had an incident at the range one day as we were conducting tracking drills with the vehicle mounted weapon against moving targets.

As one of the gunners was behind the sight on the I-TOW, he kept looking up from the viewing portal of the primary aiming optic. Suddenly, he jumped up and hopped out of the V-150, yelling something to his fellow students in one of the local languages, and several of them started running downrange.

He had seen a gazelle in the brush and now, he and several of his companions were chasing it. They caught it and quickly butchered it right there on the range. It was to be their dinner after training that day once they returned to their billets. I simply looked at my Teammates and said, just when you forget you're in Africa, something like this happens.

At the end of our three iterations of training, we had a live-fire demonstration that was attended by the US Ambassador to Botswana, the Botswanan Commander of the BDF, and other Host Nation VIP attendees. Prior to the demonstration, I was speaking with the BDF Commanding General and told him that I felt the training had gone extremely well and that he could be proud of his soldiers. He looked at me and as serious as he could be, told me that if the gunners hit the targets, then the training had been good. If they missed the targets, then the training was not good.

The pressure was on, that's for sure. Thankfully, we had direct hits on every demonstration fire mission. For the demonstration we actually used live HE or actual High Explosive missiles, not the training rounds that simply marked the target. As the rounds hit the various old vehicles downrange, there was a nice explosion and loud boom. When it all was finished, the General told me that the training had been good. He also relayed that he was ready and now able to start destroying the SADF tanks and other armored vehicles should they be foolish enough to cross the border again. There was no doubt what these new weapons were meant to be used for, at least in the minds of the BDF and its Commander.

Ocean Venture, Antigua, West Indies, SFODA-594 (MFF)

The mission to Antigua was a much different mission than any I had been on before. Instead of the tundra of Africa or the deserts of the Middle East, we found ourselves on an island surrounded by crystal clear blue ocean in the middle of a tropical paradise in so many ways.

We would train the Antigua-Barbados Defense Forces (ABDF) in light infantry and general soldier skills as a first phase of the operation. For the second phase of the mission, we would move into a Counter-Narcotic role and work with our counterparts to interrupt a sizable segment of the illicit drug and cocaine supply route targeting the United States.

One of the elements of the operation required my team to set up a drop zone for a company from the 75th Ranger Regiment that was parachuting in as direct support. The mission was also unique in that the host nation soldiers only worked during the day, Monday through Friday. With the exception of the actual Counter Narcotic mission and the insertion of the Ranger Company at the end of the deployment, we were able to enjoy some tourist time while in this vacation destination for so many normal people.

Swimming with Barracudas

While in Antigua, we were billeted in a US Navy facility and had a barracks building that was right next to the ocean. Just outside the doors was the beach and we could hit the water within a couple dozen steps. Swimming near the island coast, or in one of the bays that surrounded the peninsula on which we were conducting the major portion of our courses of instruction, was a blast. The water was warm and clear, and an off-shore barrier of coral reefs kept the waves and current from being any problems.

However, if you ventured out near the coral reef, there were a lot of fish to see and to swim with if you wanted to try. As I was swimming very close to a Grouper about 15 feet beneath the surface and near the coral, I saw a large Barracuda swim between where the Grouper was swimming and where I was swimming. I also noticed the teeth and understood that I was being told to stay away from the fish by the one that could eat them for dinner, or take a bite out of me.

I did manage to find several large Conk Shellfish while diving and brought them to shore for our fish cookout later on a weekend. I was able to get the Conk, or Conch mollusks out of the shells without breaking them by first freezing them, and then slowly pulling them out after they thawed. I brought several of the large shells home and used them as decorations since they were very pretty once cleaned up.

Pirate Castles

I'm not sure if these were real pirate castles, but they were real fortress ruins. These ruins were in several areas around the coast and we visited not only the ruins, but several resorts as well. The beaches and service by the resort staff extended all the way to the water's edge in most cases.

A Navy pickup truck and a Team Warrant Officer who didn't mind driving on the left side of the road were our tickets to sightseeing and relaxation during our down time on the weekends.

As luck would have it, one weekend we were off and able to get out and about happened to be the weekend of an annual sailboat race. There were many boats and people on the island to be at the races and much of the beach area was transformed into one long party zone. Many boats of all sizes would dock as close to the beach as possible and people would walk out to the boats through the water to visit.

The various resorts and hotels had moved serving stations down to the beach along with bands or DJs. You could walk from one party site to another within a few minutes of each other. The music and water, along with any type of drink you could imagine provided some needed relief from the stress and physical exhaustion of the military requirements of the mission. It was unlike any deployment I had been on before, or would ever go on in the future.

The 5th Became the 3rd, SFODA-594 Became SFODA-334

Operations and Intelligence School, or O&I, was a hard, but really great course to attend. I was glad to get a chance to attend and change my job description, my MOS from 18C (Engineer) to 18F (Intelligence) as I had been working as an 18F for a while anyway. I enjoyed the 18F job much more and was more experienced with it as some of the Intelligence aspects were even part of my old positions while still serving in the Artillery.

During my time in O&I, which was about 12-weeks, the 5th Special Forces Group moved from Ft Bragg, NC to Ft Campbell, KY. Both the 1st and 2nd Battalions of 5th SFG physically moved to KY, while the 3rd BN, of which I was a member, remained at Ft Bragg and became the 1st Battalion of the newly reformed 3rd Special Forces Group.

So, I started O&I as a member of SFODA-594, C Company, 3rd BN, 5th SFG(A), and ended O&I as a member of SFODA-334, C Company, 1st BN, 3rd SFG(A).

Besides the unit designation, a lot of other things changed. Suddenly, we went from being part of a fully functional and funded SOF unit, to being something new, lacking two Battalions and many personnel in support, as well as being low (always) on funding.

It became a long-running joke that the people in 3rd Group were almost expected to fund their own training exercises. In fact, we had some t-shirts printed up with the 3rd Group logo and a banner that read, “Dirty Deeds, Done Dirt Cheap” which was almost immediately banned in the Group area by the Commander.

So, as I finished O&I and returned to my old, new unit, I found out that my Team Sergeant was not happy with the situation and somehow got assigned to recruiting duty. I then became the new Team Sergeant and started putting the new skills I had learned during O&I into action as we were also told at the same time to get ready to go to war.

While I was in O&I, Desert Shield had kicked off and 5th Group had immediately sent some Detachments to the Middle East. Several of the 5th Group guys in our O&I class got pulled and had to return to their Detachments. Meanwhile, as one of the old 3/5 SFG guys, now one of the 1/3 SFG guys, we assumed that this would be another war we'd miss.

But soon, we were told to get ready to deploy for combat. As the new Team Sergeant, I had to help the Detachment weather the frustrations felt as we were soon getting yanked around with warnings to get ready, followed by being told we weren't going. This happened several times before we got the final world that we were actually going to deploy. We got the final word about two or three weeks before we went. A new Group, with only one Battalion and lacking equipment, personnel, and support. But we had a Battalion Commander who knew some of the right people and suddenly, to the surprise of the Desert Storm Commander, 4-Star General Schwarzkopf, we were in Kuwait and raring to go. The training was about to get real.

Desert Storm – The 3rd Group History

Soon after I returned home from Desert Storm, I wrote a lesson outline for military education that covered all the major actions taken by my unit during the conflict. It was a short conflict, but much was accomplished and many were quite heroic and unheralded for their outstanding feats. I've included this interesting insight into some history that is often forgotten after having been eclipsed by the extended wars in the Middle East as our nation began its Global War on Terrorism after the events of September 11, 2001.

This is more than my story. It is the story of the accomplishments of the men who were assigned to the 3rd SFG(A) and what was actually conducted by what was then, a brand-new unit that was ordered into combat. I had the honor and privilege of being a young Team Sergeant on a Special Forces Operational Detachment-Alpha (SFODA-334) during the campaign. I was assigned to Charlie Company of the 1st Battalion.

On 02 August 1990, the Iraqi military invaded Kuwait. The drive swept through Kuwait City and pushed the Kuwaiti Army all the way into Saudi Arabia. The United States quickly started deploying units into Saudi Arabia on what was called Operation Desert Shield.

3rd Special Forces Group (Airborne) had three teams deploy almost immediately, along with the 5th SFG(A), into Saudi Arabia to work primarily as liaison between the US VII Corps units and the Special Operations Central Command (SOCCENT).

3rd SFG(A) was still being filled as a new Special Forces Group and at the time consisted of a group headquarters and one battalion, the 1st BN. Our unit was first notified that the remaining elements of the Group would deploy to Saudi around September, 1990. The 5th SFG(A), which was on the ground and working with SOCCENT had determined the Special

Operations element could use another battalion of SFODAs and another headquarters element for missions and planning.

This first notification was soon followed by the message that the rest of 3rd Group would not deploy, at least not yet. This on-again, off-again notification continued and switched about every 2 to 3 weeks for a couple of months. Our BN had many other peacetime operations that were being conducted and demanded the concentration of our people, so these training missions were given priority over the possibility of deploying to Saudi anytime soon. It was our opinion that the training we were conducting, no matter what it was, would help prepare us, as a unit, for any combat mission we might get if we did deploy.

We were a new Group, but the majority of the personnel at the team level and in command positions were experienced Special Forces soldiers. Also, most of us had been together as a unit for a long time already as members of the 3d BN, 5th SFG(A). It was that unit which was used as the nucleus when 3rd SFG(A) was brought on line in June of 1990. Because of this, the majority of the personnel were highly trained in various combat and survival operations in a desert environment.
As the Christmas season approached, we were fairly confident that we would not be leaving soon, if at all. The commander authorized most of the BN to take Christmas holiday vacation. That's when the rumors started that we were to deploy right away. A few of the key personnel returned early and began planning for a deployment to Saudi with the exact nature of the mission still being an unknown part of the equation.

As we entered the new year, everyone now was sure we were going and bags were being packed.

On January 17 1991, we were in the company area having a Team Sergeant's and Team leader's meeting with the Company Commander. During the meeting someone called and said to turn on the TV or radio. When we did we saw pictures of Baghdad, Iraq getting bombed by the coalition Air Forces. No more guessing games, Operation Desert Storm had begun.

The next few days passed in a maddening bustle of activity while everyone finished getting the final things done. Then on 20 January 1991, we were officially alerted that we were to deploy to Saudi Arabia in support of Desert Storm. The advance party left Fort Bragg on 24 January and the main body departed Ft. Bragg on the 27th of January.

Once we arrived in Saudi, we stayed at the King Fahd International Airport (KFIA). It was still under construction and had the runways and some of the buildings completed. The main terminal had not been completed but 5th SFG(A) was living and working out of the basement portion.

The 3rd SFG(A) moved into the second level of an adjoining parking garage. This is where we lived, trained and planned our operations for the next three weeks. The training continually changed focus as our missions were adjusted and narrowed by the higher headquarters.

The 3rd SFG(A) began the operation with a dual tasking. Our missions were tied into two main portions of the conflict. First was the mission within Iraq itself. This was broken down into two sub-missions. The first being to conduct reconnaissance missions of the enemy in the US VII Corps sector of Iraq to determine the Iraqi's intent and provide an early warning to the VII Corps. The second submission was to conduct additional reconnaissance missions into southeastern Iraq to determine whether the terrain would support movement of the armored vehicles in the projected area of operations for VII Corps.

The second proposed mission was for inside the country of Kuwait. This mission statement basically directed the 3rd SFG(A) to prepare to coordinate with the Kuwaiti Army General Headquarters and establish a combined Special Operations Headquarters to provide a connection between units of the Special Operation Forces, Kuwaiti resistance forces and allied conventional forces. Also, to assist in the clearing and securing of Kuwait City.

The purpose of the dual missions was to support the coalition forces Desert Storm plan for the defense of Saudi Arabia and for the recapture and restructuring of Kuwait and Kuwait City.

Operation Desert Storm had an annex, or sub-operation plan that was titled "Operation Urban Freedom". This OPLAN began implementing actions that were to be conducted around Kuwait City to ensure the orderly forward deployment of the Special Operations Central Command forces. Also, this plan would allow a smooth flow of the coalition forces, which were primarily U.S. Marines and Kuwaiti forces into the area to assist in the liberation of Kuwait City.

The commander's intent was to deploy SOCCENT forces forward to the vicinity of Kuwait City as the conventional coalition force secured Kuwait up to the 6th ring road of Kuwait City. Kuwait City is surrounded by circular roads that encompass the city center which is bordered by the gulf on the east and north edges.

Once the conventional forces had cleared up to the 6th Ring Road, a joint task force composed of Kuwaiti forces and the SOCCENT forces were to clear and secure the city.

The U.S. task force personnel would assist and advise the Kuwaiti forces throughout the liberation of Kuwait City and would occupy the U.S. Embassy in a coordinated effort to occupy several embassies simultaneously, by different country's Special Operations forces.
Specific 3rd SFG(A) missions that were proposed in support of the mission in Iraq included:

Direct Action missions; Rear area lines of communication, Main supply routes, command and control facilities, and mobile targets not easily destroyed by air assets.

Strategic Reconnaissance missions; Hidden airfields, high value enemy strategic targets, command and control nodes, and early warning.

Combat Search and Rescue missions, Primarily in southern Iraq.

Foreign Internal Defense missions, In Saudi Arabia.

Specific 3rd SFG(A) missions in support of the mission in Kuwait and Kuwait City included:

On order, forward deploy forces to Kuwait International Airport and establish a forward operational base.

Be prepared to link up with resistance forces to conduct combined operations.

Be prepared to assist resistance in utilization of Close Air Support, (CAS).
Be prepared to support link up of resistance forces with coalition forces.
Be prepared to secure designated key facilities in Kuwait City in coordination with Coalition and/or U.S. forces.

Be prepared to assist in the conduct of Civil Affairs and Psychological Operations beyond the capabilities of other forces.

Be prepared to conduct Combat Search and Rescue.

Be prepared to accept operational control of a Kuwaiti Special Forces Battalion.

On order, secure and occupy the U.S. Embassy in Kuwait City.

And, on order, assume command, less OPCON, of 4/17 Cavalry.

These were all the scenarios and missions the commander of Special Operations Central Command had envisioned could be given to the 3rd SFG(A). This is what we began to prepare and plan for during the initial few days after arriving in Saudi. The following, are highlights of what the 3rd SFG(A) actually conducted once the missions began:

The 3rd SFG(A) conducted the initial two missions, however the focus was narrowed quite a bit. First the Reconnaissance teams went deep behind enemy lines inside Iraq. This was in support of the US VII Corps. The Reconnaissance teams' missions were two-fold. First, they were to find the Iraqi Republican Guard Army and their respective divisions. Secondly, when the attack kicked off in February, they were to determine whether the Republican Guards Divisions were going to come south and meet the VII Corps attack, if they were going to head west to Baghdad, or whether they were going to remain in position.

The second mission, inside Kuwait, was also narrowed considerably. This mission was restated directing 3rd SFG(A) to seize, clear, and hold the U.S. Embassy until relieved.

To accomplish the first mission, several teams, to include my Team, began to experiment with different types of desert hide sites and techniques to remain hidden. Due to the variance of terrain many ideas were tested. Some of these included using plastic pipe covered by various types of cloth material, sandbags, digging holes, and hiding in whatever terrain was available.

Three teams actually went inside Iraq. Two teams stayed until their mission was complete and one had to be extracted almost immediately due to an enemy tank unit in the area. There was no loss of personnel or equipment on these missions. They were infilled and exfilled by helicopters that came from either US Army or Air Force Special Operations Assets.

The mission in Kuwait City began with the Commander, LTC Frank J. Toney reorganizing the battalion into a special task force. He divided the three companies of the battalion into an assault force, a security force, and a support force. There was very little time to get ready. We had less than one week from receipt of the mission to receive required intelligence, develop an assault plan, conduct rehearsals and then, actually launch on the mission.

The rehearsals took place at King Fahd International Airport area. Then the unit moved by helicopter to the Kuwait

International Airport (KIA). When the force landed at KIA there was little to no resistance. The force stayed at KIA for the night, then launched for the US embassy by wheel and helicopter the next morning. My Team participated in this as our initial operation. This was 28 February 1991. The US embassy was cleared, and then was turned over to the U.S. ambassador on the next day.

During this portion of the operation, the lack of accurate and timely intelligence was a major concern of the force. We had been told to expect to find anything from a reinforced mechanized infantry company with chemical and conventional munitions, to finding no threat at all. Because of this, the task force went in wearing chemical protective suits that were very hot and could easily have created heat injuries. The leaders were aware of this but were trying to protect their soldiers by lowering the number of possible casualties for any situation. Also, the water, ammunition, and anti-tank weapons the men were carrying created a rucksack that weighed more than 120 pounds or 55-60 KG and a combat vest weighing around 70 pounds or 30 KG. Everyone was also wearing flak jackets or body armor, helmets, and carrying protective masks. This made the individual soldier's load very heavy and would have been very difficult to maneuver with had the enemy chose to fight.

Due to smoke from the oil wells that were burning all around Kuwait, visibility was extremely limited. The wind shifted after the US Embassy had been entered and within an hour it was dark as midnight, even though it was still midday. This made the job of securing the Embassy slightly more challenging and the shadows created problems where there should not have been any. The embassy compound was not defended nor had it any enemy troops inside. Reports received about booby traps, left from both the Iraqi's and the US personnel as they evacuated the embassy, coupled with the urgency for speed in clearing the facilities, created an environment that allowed some demolition and searching operations to take place that caused significant damage to portions of the compound. However, the commander said he needed to be sure it was safe for embassy personnel to return. Upon completion of the securing of the embassy, our mission shifted to embassy guard and VIP security. At the same time, we began planning for reestablishing some kind of order in Kuwait City.

As most people know, the ground war only lasted a "magic" 100 hours, or 4 days. This was not expected or planned for and there were no plans conceived yet for the reorganization and rebuilding of the infrastructure of Kuwait City.

We had several challenges, even though it wasn't part of our Mission Essential Task List (METL), or part of our directed mission. Suddenly we needed to get water and electricity back into the embassy and to help make possible the eventual return of these products to the entire city. We maintained and helped control food lines and attempted to ensure human rights violations didn't occur. Several times, teams were sent to police stations to prevent or stop atrocities that were being committed against suspected Iraqi stay behinds, sympathizers, and Palestinians. Palestinians composed the major part of the work force in Kuwait and many belonged to, or were sympathetic with the Palestinian Liberation Organization (PLO), but the majority of them were still loyal to Kuwait or at least were not supportive of what the Iraqi Army had done. Many Explosive Ordinance Disposal (EOD) missions were performed on stockpiles of ammo, booby traps, and land mines. This was compounded by the massive amounts of unexploded air ordinance left from the intensive bombings prior to the ground war that, was literally, scattered everywhere.

The Commander once again broke the battalion down by company and gave the mission of guarding the embassy and providing VIP security to B Company, he gave the Eastern area command to A Company and the Northern area command to C Company. The Northern area command included the mission of sending one Operational Detachment-Alpha (SFODA) north of Kuwait City near the border, to work with a Kuwaiti armor brigade. This unit was right in the middle of US VII Corps armor units. Deconflicting the problems between the Kuwaiti unit and the US unit was an ongoing problem for the Detachment Commander. The main problems came from the conventional officers not being fully aware of the customs and cultural differences. Also, since the Kuwaiti Commander kept crossing into disputed territory or would maneuver his unit into the US element's area of operations in pursuit of Iraqis, suspected or otherwise, matters were not made any easier.

The other teams in C Company and the ones in A Company set up headquarters elements alongside their Kuwaiti counterparts at the division level and sent teams to Kuwaiti military and police outposts throughout the city. Once deployed, they continued to help organize a system of control and perform Explosive Ordinance Disposal and quick reaction force missions. As munitions caches were found, or small attacks via sniping or drive by shootings would occur, we responded alongside our assigned Kuwaiti units. During the first week after securing the embassy and these other operations had begun, we conducted numerous Direct Action and information collection missions throughout the city. Some of these missions included the following:

The Kuwaiti government experimental farm, used as a major Iraqi interrogation and torture center. War crime evidence was gathered and preserved.

The Al Shaab Palace, Iraqi interrogation and torture center.

The Sief Palace, Iraqi interrogation and torture center. It produced major war crime evidence.

The Muk Harbaratt HQ's, it was utilized as the Iraqi command and control HQ's for Iraqi intelligence soldiers. It produced critical information on their military operations and war crime evidence.

Several police stations, used by Iraqis as interrogation and torture centers.

The Kuwaiti women's prison was used as an interrogation and torture center. It produced war crime evidence and documentation.

The PLO HQs provided information on command and control of terrorists and military organizations that were loyal to the Iraqis and supported them during the occupation of Kuwait City. It also produced plans, terrain models, names, passports and Counter-intelligence (CI) documentation.

The ice-skating rink was used by Iraqis as a morgue. It provided information and bodies bearing evidence of the torture techniques the Iraqis used.

All colleges and university dormitories and classrooms were used as living and working areas by Iraqis. They produced many documents, weapons and ammunition caches.

Hotels and residences were used as working and living spaces by Iraqis. Many of the hotels were burnt down. These produced some documentation, weapons and ammo.

Beach trench and bunker complexes produced many anti-aircraft and artillery weapons, small arms and ammo. This entire area was extensively mined and contained some booby traps.

My Team spent quite a few days securing several of these areas, as well as clearing several hotels as well as a large school facility. We came under fire at a beach area during one of our clearing operations, however the attack faded into the city as we responded.

As we started to bring the city under control, we were assisting the Kuwaitis more and conducting missions less. We were prepared to stay in Kuwait City for 3 to 6 months and continue with the stabilization mission. Then abruptly on 12 April 91, to the surprise of our commander and our Kuwaiti counterparts, we were ordered to redeploy to the United States. The 5th SFG(A) then stepped in and assisted the Kuwaitis we had been working with. On the 26th of April the main body of the 3rd SFG(A) returned to the United States.

This story has not been all-inclusive, but it contains all of the major documented missions that were performed by the 3rd

Special Forces Group (Airborne) during Operation Desert Storm. Thankfully, none of the men from the 3rd SFG(A) were killed during this conflict.

The Captain Who Couldn't

During Desert Storm, I was the Team Sergeant for my Detachment. I was a Sergeant First Class (SFC E7) filling both the Master Sergeant's slot as the Senior NCO, as well as the Team Warrant Officer or XO's slot. I had a Captain as the Detachment Commander and six more NCOs. Our 8-man strength was quite normal at the time, due to a lack of SF qualified NCOs throughout Special Forces.

For reasons known and unknown to me, our Battalion Commander didn't interface much with NCOs when he provided guidance and direction to the Company Commanders and Detachment Commanders. He had spent much of his career in a Ranger Battalion and brought a lot of the Ranger mentality to SF when he arrived in 5th SFG(A).

As we prepared inside Saudi for movement to, and operations inside Kuwait, the planning discussions were almost always done during officer-only meetings. It was after one of these meetings that I had my confrontation with the Detachment's Captain.

His first name was Brian and his last name's first initial was S. CPT BS (appropriate initials) had been the Team Leader for a little while but never really fit the mold. It got much worse after we deployed for war. After he returned from the latest BN CDR's meeting, he told me that he needed to talk, but wanted to speak to me privately, away from the Team. We walked around the perimeter of the Company area and then he started to spill his guts.

He first told me that the BN CDR had asked for a volunteer to have a Detachment stay behind as the Rear Detachment while the rest of the Battalion deployed into Kuwait, and that he had volunteered our Team to remain behind. This really made me mad and I told him so in plain language. I also said that he should have discussed it with me before volunteering.

It was then that he told me the truth behind his call. He said he had made a big mistake by going SF in the first place and that he was not cut out for this type of work, let alone combat. He further stated that he was afraid to go into combat and that he was ashamed for himself and disgusted that he had brought this shame to his family, etc. He was practically bawling by this point and I was totally disgusted.

He was coming apart, having a mental breakdown, right before my eyes and was almost out of control. He no longer wanted to be in SF, he no longer wanted to be in charge, he no longer wanted to be deployed. Looking back, I see now how I could have handled this incident better, but at the time I was much younger and under a considerable amount of stress myself. Instead of being sympathetic or understanding and possibly attempting to help, I just blew up.

I didn't retain much professional composure and I sure didn't treat him like any officer I had ever interacted with before. To put it mildly, I told him he was fired and to get his gear out of my Team Room and never come back, under threat of physical harm. He got his gear and departed as soon as we returned to the Team area. I told my Teammates what had happened, and I also told two of the other Detachment Commanders who were friends of mine. They also became irate over what he had done and how he had done it. I didn't know where CPT BS had gone, and frankly I didn't care. I continued to train my Detachment as usual. After about 48 hours, I was called to the Company Commander's office.

Once I got inside the CO's office, I was read the riot act. I was told by the Major that none of his Captains were cowards and how dare I treat a Captain in such a way. I argued to no avail, at least until I was ordered to stand at attention and stop speaking. My Company Sergeant Major had nothing to say, pro or con, and simply let the Major continue to dress me down for having the gall to fire one of his Captains. It was at this time that I was also told that I would not have to request a reassignment when we returned to Ft Bragg. He was going to ensure that I no longer remained in his Company as soon as we got home.

Soon after I returned to the Team Room, CPT BS slunk into the area and set his gear back down by his bunk. We didn't talk until the next day and then we only discussed the upcoming events and how we were being told to participate in them.

When the Company deployed into Kuwait for the assault to secure the US Embassy and the inner "ring roads" of Kuwait City, my Detachment was split apart and the guys joined other Teams as extra personnel. CPT BS and I remained with the Company HQ under the watchful eyes of the Major and SGM. Within a couple of days after going into Kuwait City, I got my Detachment back and we were initially assigned security duty for the top floor of a nearby hotel where VIP and high-ranking personnel would stay during their visits to the US Embassy or operational HQs we had established inside Kuwait City. CPT BS remained with the Company HQ and worked for the MAJ for most of the time we were in Kuwait City.

For most of the remaining time, I took my Detachment on multiple operations in and around the area. For one particular patrol, CPT BS returned to participate as we went to clear a major hotel that had been used as Iraqi barracks and to clear the adjoining beach front area. During this operation CPT BS apparently lost a set of PVS-7 night-vision goggles, but failed to tell anyone or to check his equipment before we left the area. Later I discovered that the Company Commander tried to initiate an investigation to determine if I had had anything to do with his loss of equipment. I didn't, but the implication really frustrated me and quite a few others in the Company.

Immediately upon redeployment to Ft Bragg, I was reassigned to the Battalion Headquarters as the Intelligence Section, or S2 NCOIC. As soon as the Company Commander left my old unit, the Group's Command Sergeant Major directed my BN CSM to send me back to the Company and back to an SFODA where I was needed more than I was needed in a Staff position.

The cowardly CPT BS? He left Special Forces as soon as we returned to Ft Bragg and got an assignment with Civil Affairs. And from what I heard, he never returned to a real SF position.

Hotel & Beach Patrol in Kuwait City

It was mid-morning, but it looked like night. The wind had been blowing the smoke from the burning oil wells back over Kuwait City and had blocked out the sun. Everything smelled like oil, even the food. My uniform was starting to turn from the tan and brown Desert Battle Uniform or DBU, into a mixture of black and gray. I wondered briefly what the oil was doing to our lungs, but couldn't dwell on it.

The Company was going out on a multi-location search for Iraqi soldiers and any associated Intelligence or caches of supplies. My Detachment was headed for a large hotel along the beach area of the north-east part of the city's center. This was a location that had been used to house Iraqi soldiers. The beach was part of a defensive perimeter that had been established to protect the city from an expected waterborne assault.

When we got to the site, the hotel had suffered obvious damage from not only bombs, but from fires that had probably been set by the Iraqis as they fled the area. I remember going through the floor with several shops for the tourists who once frequented the place while on vacation. There were tennis rackets strewn around the area that were priced at hundreds of dollars apiece. What hadn't been stolen or destroyed was ransacked and left behind.

In the basement stood a large safe. It was a free-standing safe that must have been six feet tall and as long and wide. It showed signs of damage, as if someone had tried to force it open. It was also blackened on one side where it had been in a fire that ruined part of the basement area. However, the safe had not been opened that we could tell. I wondered if the hotel management had emptied it as they ran from the Iraqis when the city was attacked, or if there was still untold treasures or cash setting inside the safe.

It was while I was in the basement area that I heard shots being fired and one of my teammates yelling that we'd started to receive fire. I ran to the rear of the hotel which overlooked the bay, and saw a couple of my guys behind cover, scanning towards the north for any attackers. I stepped out into the sunlight and started maneuvering the team towards a perimeter wall that tied into a recently created berm running toward the water.

We didn't see anyone in the area, nor did we receive any additional fire. At this point we moved our search from the hotel onto the beach complex and quickly discovered there were dozens of small anti-personnel mines laying on top of the sand, spread all across the beach. Upon closer inspection, it appeared that the mines had not been armed, at least the few in our immediate vicinity were still safe to handle. As we moved towards the trench line that had been dug, we found that the mines had only been spread in the area right behind the hotel walkways headed towards the beach.

Up the beach towards the north, I went into a storage building and found a large box that was marked as explosives. We had brought a couple of USAF Explosive Ordinance Disposal or EOD technicians with us on the patrol, so I called them forward to inspect this large box. One of the EOD guys moved towards the box and looked around for a few seconds and then just grabbed the lid and opened it up. I couldn't believe what I'd just seen and told him so. I said that anyone could have just opened the box without checking first. I then told him and his partner to return to the vehicles and wait there. The box was filled with ammunition and some hand grenades, along with some more anti-personnel mines. Thankfully, it had not been booby-trapped.

As we moved further up the beach head, we came to a dug-in ZPU-4 pointed towards the bay waters. This was a four-barreled heavy machine gun normally used as an anti-aircraft weapon. However, this had been set up in preparation of fighting against ships which they had been convinced were going to storm the city. All of the Iraqi defenses were focused towards the water with little effort having been made for thoughts of defending an attack from the land routes.

This gun was still loaded and after a quick inspection by our weapons NCO, it was deemed safe for use. Several of us took turns firing the gun out towards the water. Then, after some additional thought, we decided that since we didn't know exactly what was out on the water, we should stop shooting. We didn't want to accidentally shoot up our Navy brethren, or worse, get someone to shoot back at us.

This patrol was followed over the next few days with more that were similar, but in different types of locations. We raided a known PLO HQ site and captured a lot of written Intelligence documents, maps, and photos. We raided a school dormitory that had housed Iraqi troops and recovered a lot of small arms and ammunition. We even went into a suspected terrorist or PLO cell and found a huge sand-table that had been built, along with a still hot pot of tea and a couple of Iraqi uniforms laying on the floor. These enemy soldiers must have heard us coming and changed clothes on the way out to blend in with the rest of the locals.

And almost every day, we were covered with an oily haze when we got back to our base camp. The wind would shift and it would go from sunny and bright, to dark and nasty. Everything became gritty as the oil would attract the dust and sand, and everything was covered with a little oil from the smoke. But the oil wasn't the only evidence of the war's destructive results.

The Highway of Death

As the US Coalition, which was mainly US, started to pound the Iraqis across Kuwait, the ones inside Kuwait City started running like stampeding donkeys for the border to Iraq. They took every military and civilian vehicle they could find and headed out on the one big highway back home. Too bad for them, the highway was patrolled by the American A-10 Warthogs and AH-64 Apache helicopters. Once the Iraqis began their rush to retreat, US fighters stopped a convoy carrying thousands of Iraqis, weapons, and stolen loot that was about six-miles long and spanned all lanes, bumper-to-bumper of a six-lane highway. As the front of the convoy was stopped by rocket and bomb attacks, the rear of the convoy was hit as well, preventing anyone from turning around.

The sand of the desert as well as the rough terrain would have made it almost impossible for many of the vehicles to have departed the highway and escaped cross-country. However, after the convoy of vehicles had been effectively trapped in place, the jets moved out and the Warthogs and Apaches moved in. It was as close to shooting fish in a barrel as you can imagine, without the fish.

As I later traveled the Highway of Death, I saw where in the panic to escape, cars that were found just off the road had gotten stuck and had been run over by trucks, which in turn had been run over by armored vehicles as they tried to flee from the rockets and missiles raining down from above.

I also spoke to a couple of A-10 pilots a few days after the event and they explained that it was like a deadly game. One A-10 would flip a tank over with its rockets and another would flip it back over during a second run. I also saw an Iraqi 155mm self-propelled howitzer that had been hit with ordinance so hot that the only thing recognizable was the muzzle at the end of the cannon, and a couple of road wheels. All the metal, which is mostly aluminum, had melted into the sand.

Of course, there were hundreds, if not thousands of dead bodies. Many were burned, but many were also crushed beneath the piles and rows of destroyed vehicles. The smell was horrible and within a few days, the entire area was off-limits to the many soldiers and civilians who had ventured into the area after the fighting was over. Their units were essentially conducting sightseeing tours. This was quickly terminated by the area commander as our units continued to clear and control the aftermath of the US assault into Kuwait City.

During the reconstruction and assistance period, some well-intended personnel tried to cast doubt over the assault operation by implying that the amount of force and destruction was avoidable and even unnecessary. But no punishment was ever offered that I was aware of for the annihilation of the Iraqis fleeing the wrath of judgement brought to them by the US military along route 80 headed north to Baghdad.

Recovery of USMC Property

My Detachment was on a patrol in the area to the north of Kuwait City to search a bunker and trench position along a ridgeline that faced southward into the city's open areas between the north-running highway and the coast. As we drove to the mission site, we noticed a USMC Humvee (HMMWV) parked among some of the destroyed Iraqi vehicles still littering the side of the roads in the area. I assumed that it was just some souvenir hunters or people out looking to see what they could see and thought no more about it.

Once at the bunker and trench line area, we completed our mission, finding nothing of any real value. As we returned to our base in Kuwait City near the US Embassy, I noticed the USMC vehicle was still parked in the same place. This was not something that was normal, so we stopped to investigate.

The Humvee was empty of any personnel, but was full of gear and equipment. It also had a hole blown through the rear floor as if it had driven over a small mine. Inside the vehicle was someone's load bearing gear, which included a personal sheath knife and some pictures of an American family, a wife and a couple of kids, in one of the ammo pouches.

In the back of the vehicle was a complete secure radio set, and the encryption code was still loaded into the communications gear. There was also what seemed to be a nearly brand-new M-60 machine gun, spare barrel, and ammunition. Obviously, something was wrong with this picture. We made a search of the area to try to locate who might have driven the vehicle to this location but couldn't find anyone. There was some blood in the vehicle, so I thought that probably whoever had been in the vehicle had been taken to get treatment by someone else. Still, that didn't explain why the vehicle was there, or why it still had all the gear and classified commo equipment present.

It was nearing the end of daylight, so we recovered the gear and made sure the vehicle itself was as secure as possible and returned to base. Once I reported the incident to the chain of command, we were able to contact the Marines in the area and alert them to what we had found and where we were keeping the gear we recovered.

Soon, we were visited by a Marine Major. He was the one who owned the web gear, and whose family we had seen in the picture. The Humvee was his vehicle, which had been stolen from their compound and was lost the day before. He didn't have a story about how or why, or even much more than a "thanks" for recovering his personal and military gear. I always felt that there was more to this whole story than met the eye, because we never received any type of formal recognition for finding and recovering the gear, let alone any further information about the outcome. In hindsight, I've often thought that we should have just kept the equipment and increased our own unit's store of extra supplies, especially with the M-60 machine gun.

Reorganization, Policing, and Civil Affairs

Our mission set changed quickly. For days, we had been searching for enemy and expecting to engage in fire fights on every new excursion into the former strongholds of the Iraqis. Now, we had become the police and the answer for all local problems.

My Detachment continued to be under the direct control of the Company Commander after the issue we had gone through with the Captain. The Captain worked in the Company HQ and I, along with several of my Detachment members worked various taskings in support of the changing mission focus. I had initially begun to collect and organize a data base of the Kuwaiti military unit we were supplementing and then passed it along to the B-Team's Intelligence NCO. I started working directly with one of the Kuwaiti staff officers and organized distribution of logistical resources to residences and businesses that had contacted the Coalition Forces for assistance. Water was the most critical resource as the desalination facilities in Kuwait City had been destroyed by the Iraqis at the end of their occupation. Many water truck deliveries had to be scheduled and some needed to be monitored and provided with security.

Additionally, several of the ammunition caches in the city were now in danger of being looted and had to either be secured, or destroyed. One of the more challenging functions was to respond to acts of retribution by the Kuwaitis against the large population of non-Kuwaiti personnel in the area who had comprised the large part of their work force before the invasion. Apparently, a lot of the PLO affiliated or non-local population were sympathetic towards the Iraqis, or at least didn't suffer as much from Iraqi control as did Kuwaiti personnel who could not escape prior to the city being captured.

As the situation continued to change, our unit was flexible enough to shift gears and get on top of whatever requirements came our way. Just as we were getting used to the transition from warrior invader to welcomed liberator and provider, we got our orders to return to Ft Bragg as the 5th Special Forces Group assumed all our duties and took over all of our taskings.

Suddenly for us, the war was over, or at least our part of it had been completed. We were going home.

Getting ready to return to the States was a much bigger hassle than it was leaving. As the containers were being packed and the gear being cleaned, inspected, and made ready, I was notified, along with three or four other senior NCOs that we had been selected to attend the next iteration of the Advanced Non-Commissioned Officers Course (ANCOC), a career-progression requirement for promotion to Master Sergeant. We were to depart a couple of days before the main body would fly home. We flew back on a commercial airliner that was also bringing another unit back to Ft Bragg. As we loaded onto the aircraft, the crew put us into the first-class seats and we promptly feel asleep. At some point the unit loaded and when I woke up I saw the Command element of the unit setting with us in the first-class cabin, no doubt wondering why these guys in dirty DBUs with no rank or unit identification patches were in the first-class section with them. I smiled at the thought and went back to sleep.

Just before we were about to take off, one of the crew from the flight deck asked me if I wanted to ride in the cockpit as we flew out of the country, so I watched us leave the ground and fly away. I felt the weight of the whole experience disappear like the desert was disappearing below us. Soon it grew dark and I returned to my seat below and went back to sleep. I had a lot of sleep to catch up on, and this was a perfect opportunity.

Eventually, we landed at Pope AFB, which was collocated with Ft Bragg, NC. As I got off the aircraft and avoided all the welcome home hysteria awaiting the unit whose aircraft I'd shared, I found my own family waiting on me. I was still in sort of a fog, but I knew then that I was really home. The next day or two remains a blur. I just know that soon, I was in school and trying to focus on academics while my mind was still in the sands around Kuwait City. To say that I was not in the right frame of mind for an important school for the next nine-weeks was an understatement. However, I managed to make it through and even finished in the top 10% of the class, earning a spot on the Commandant's List for honors.

When I returned to my unit I reported in to the Battalion Intelligence office, or the S2 as the new NCOIC. Having just completed ANCOC gave me not only some very fresh insight into the Intelligence requirements of the BN, but to the administrative processes of the BN's security and clearance requirements as well. I soon discovered that the backlog of security investigations for clearances was huge given the actions of our recent deployment to Desert Storm, among other things.

As it turned out, I had an exceptional S2 Officer or OIC, who was an actual Military Intelligence (MI) branched officer who was really up on all the administrative aspects of the S2 shop regarding clearances and security background investigations. Between him and an E5 Military Intelligence trained NCO I had working as my S2 clerk, we started clearing up the admin backlog. During this time, I focused on the Intelligence side of the office and even managed to get a second S2 clerk assigned to the S2 section. The new clerk was a PV2, right out of MI training and Airborne School. Using him and his access to the MI Detachment's classified systems, along with my input from Open Source Intel (OSINT) and experience in our operational areas, I developed a running Intel notebook and briefing that I presented during each BN level Command & Staff meeting. I wasn't sure if this was a real significant creation, but it was soon a very popular part of the briefing, and especially so for the BN Executive Officer (XO) the 2nd in Command for the BN. Several years later, the XO would become my BN CDR and a strong advocate for me as I became the leader for the Sniper Detachment in 10th SFG(A) at Ft Carson, CO.

During my time at S2, I was granted a Top Secret clearance and this came in handy not only for security requirements the rest of my military career, but carried over into my contractor career where having access to the highest level of information became harder to get. Additionally, I attended another school, the Advanced Special Operations Techniques (ASOT) course. This was a course that mirrored much of the CIA type of activity for developing Intelligence networks with non-conventional and civilian personnel in foreign countries. These skills would come in handy and were put to use in Bosnia years later, as well as other locations.

At about the nine-month mark of an expected 12-month tour in the S2 shop, I was offered the opportunity to return to my old Company since by now, the former Company Commander had rotated out and been replaced. I jumped at the chance to go back to the Company, but once I got there I found out that instead of my old Detachment, I was going to be assigned to SFODA-336, a Scout Swim Detachment.

The Company SGM greeted me with a story that I hated to hear. First, he let me know that my old Detachment was keeping the Master Sergeant as the Team Sergeant after all. Originally, he was going to be leaving for another assignment, but that got changed at the last minute. That's what I was told anyway. That's when the SGM told me about the leadership on my new Detachment.

The Team Leader and the Team Sergeant were both African-Americans. The CPT was a Washington, DC area raised intellect, and the Master Sergeant was a Texas cowboy. The SGM was of Mexican heritage and I always heard he didn't care for blacks, but he never said anything to me that would indicate those feelings. All he told me was that both the CPT and the MSG were light in leadership skills and that they needed some assistance in that department. Later on, I would face a similar challenge again as I signed in to another new unit and was given a Detachment I didn't expect.

This Detachment was just a little out of control. I knew most of the guys on the Team from Desert Storm, but both the Team Leader and Team Sergeant were new. There wasn't an XO on the team, so I filled both the 18F Intel and the Warrant's duties as the assistant Operation's Sergeant. It wasn't a Military Free Fall Detachment, but it also wasn't a Staff job either. I was back on an A-Team and that was alright by me.

War Games and My Son, SFODA-336, Summer of 1992

We had something called training cycles and these provided periods where we sent people to schools, conducted actual operations, and conducted local training. During a local training exercise, my Detachment, along with a couple of others in the Company, decided we'd do some Force-on-Force operations at a nearby training area on Camp Mackall. One of the missions was a two-day event with a lot of patrols and a raid being conducted on a static position. My Detachment was going to be manning the static position, a two-story building inside a basecamp set up out in the woods. There were a few rockets and a bunker or two set up in the area and all were great places for targets or to practice small unit tactics in the offense or defense. I decided that this would be a great time to bring my eight-year-old son Josh to the field and let him work as a Detachment member in training.

I brought him to the Team Room at the Company area and had him attend our pre-mission planning and assigned him duties he would be responsible for during the operation. He learned everyone's names and what their jobs were going to be, and also how he would be included in the missions, from guard duty, to patrols, to defending against attacks.

We had decided to use blanks and the Multiple Integrated Laser Engagement System (MILES) to add realism and positive targeting for our direct-contact fire fights. When I signed my M16 rifle out from the Arms Room, I also signed out a second one for Josh to carry and use.

It was a great experience for my son, but it also provided the members of my Team an opportunity to practice their skills at teaching someone the basic requirements of security and combat tactics. Teaching an eight-year-old English-speaking boy was close in difficulty to teaching foreign soldiers who spoke a different language.

During one of the assaults against our position in the basecamp and building, Josh was the M60 machine gunner and was pretty good at aiming while ripping off a belt of 7.62 blanks and pinging the attackers with the MILES laser. Of course, I prevented him from becoming a casualty by not having him wear any of the MILES sensors. After all, I was his dad and I needed to protect my son.

He also got to learn how to use our night-vision goggles, the PVS-7, which he wore during his stint at guard duty through the night. He also enjoyed the ride from Ft Bragg to Camp Mackall. Riding in the Detachment's Desert Humvee (High Mobility Multipurpose Wheeled Vehicle – HMMWV) or what we called the "Dumvee" was one of the highlights of his visit. These vehicles were loud, rough riding, hot, and smelled like downrange. What's not to love about that when you're eight?

After the operation was over and the gear and weapons had all been cleaned and accounted for, we got to go home and plan for the next big adventure. As things turned out, that was the only time Josh ever went downrange with me again. Between deployments and having my kids grow up while I was running a fast-paced track as a Green Beret, made the time disappear too quick. But we had at least one big "Hoo-Yaa" and it made a mark that would never fade away, at least not completely.

I ended up staying with SFODA-336 for about six-months and we deployed several times. One of the deployments was to support Counter-Narcotic operations on the southern border. The SF Group was doing a lot of work across the entire southwestern region during that period. We conducted several missions for what was then called Joint Task Force Six, or JTF-6, and we worked with local law enforcement and the Border Patrol. Being active-duty military, we were limited in what we could do legally, so we had to stay strictly in support of the sworn law enforcement officers. It was interesting work, but at the same time, it got a bit old pretty quick.

Still, we made the best of it and actually got a lot of good training in while we were deployed along the border in the southern most areas of Ft Bliss, TX. Besides normal desert operations, we also did a lot of Scout Swim training, to include Helocast operations, long surface swims, and other types of waterborne training in a couple of very large lakes in the area.

Soon enough, we were back at Ft Bragg from what was to be my last deployment with SFODA-336. The 3rd Battalion of 3rd SFG was being assembled, or as we called it, stood up. The 3rd Group CSM called me and asked if I wanted to go back to a Military Free Fall Detachment, and of course I agreed. He made it happen, and I was assigned as the 18F Intel Sergeant to the newly formed SFODA-384.

SFODA-384 of B Company, 3rd Battalion, 3rd SFG(A). We were soon calling ourselves the Free Falling Farm Animals. This Detachment quickly became my new home. Soon after being assigned to my new Detachment we got ready to deploy, and this time I was headed to Zimbabwe.

Lancer Jaguar & Falcon Cherokee, Harare, Zimbabwe 93 & 94, SFODA-384

While I was assigned to my latest MFF Team, I started out as the Intel NCO and later became the Operations NCO or Team Sergeant. During my almost two-year tour, we deployed to Zimbabwe twice. Both times we based our operations from Harare and worked primarily with the Zimbabwean version of the US Army's Parachute School. The Zimbabwean Parachute Jumping Instructors, or PJI unit members were all Zim Air Force by branch, but most had served as soldiers in the Army or other types of paramilitary units during their earlier years.

The PJI unit was considered an elite unit and was also the unit providing the Zimbabwean military's skydiving team. They instructed the Zim Army parachutists and the Zim Special Air Service members on basic and advanced parachuting techniques. Additionally, they performed the duties of Jumpmaster for all Zimbabwean airborne operations.

The Zimbabwean military was a unique mixture of men. Having went through a civil war in the late 1970s between the Rhodesian government and African nationalist groups, a black-majority government was formed in 1980 and the country was internationally recognized as the independent Republic of Zimbabwe. Several of the PJI members and many of the 1st Zimbabwe Parachute Group were former Selous Scouts. The Selous Scouts were a Special Forces type of unit that fought for the Rhodesian Army during the civil war, or Bush War as it was called.

Likewise, many of the unit's members had also served with the various African nationalist groups that had fought against the Rhodesian Army. After Zimbabwean independence had been established, the former-warring factions mostly came together for the good of the nation and for many years, Zimbabwe prospered until the corrupt government of the Prime Minister, Robert Mugabe who ruled from 1980 until 2017, brought it all down.

Even with the extraordinary unification between the former enemies in what was the newly formed Zimbabwe Defense

Forces (ZDF), the ZNA (Zim National Army) and the AFZ (Air Force of Zim), there were still issues.

I once talked to one of the parachute group Corporals who was older than most soldiers holding like rank. I asked why he was only a Corporal when it was obvious that he was very experienced and knowledgeable. He simply stated that he was from the wrong tribe to ever be promoted above his present rank. Not only was tribalism obvious in their politics, but also in the military structure. Tribalism, paired with which side of the war they had fought for during the Rhodesian Bush War, played heavily towards assignments and promotions.

Zimbabwe's Tracking Combat Unit, Kariba TCU, SFODA-384

During the first deployment into Zimbabwe, my Detachment not only trained the Zimbabwe military, we also received some specialized training from the ZDF. The training was called Combat Tracking and was provided by the Tracking Combat Unit near the town of Kariba in northern Zimbabwe on the border with Zambia.

The school was the site of a former Rhodesian military unit that was instrumental in the history of both the Selous Scouts and the Grey Scouts. The course was a very difficult course, not so much for the instruction we received, but for the physical aspects of the training. Over a week, members of my Detachment were taught how to read signs of people as they moved through the jungle and tundra of the area around Lake Kariba and the game preserves in the border region between Zimbabwe and Zambia. Much of the combat tracking was done while quickly walking or even jogging across various terrain with a full combat load.

During one movement, I was leading my group of trackers down a trail following "spore" which is what the Zim instructors would call signs of the one being tracked. As we rounded a corner with tall vegetation on either side of the trail, I saw an antelope dash across several meters to my front. I didn't think much about it until a couple of seconds later, a lioness bolted out of the jungle less than 15 feet to my front and stopped momentarily, looking at me as I also stopped in my tracks. Within a second, the lioness continued across the trail and disappeared into the foliage apparently continuing to pursue the unlucky antelope.

Our Zim tracker instructor told me that this was an area where many lions had gathered because it was relatively safe from poachers and there was a lot of game for them to eat. Later as we set up camp for our overnight halt, we maintained a couple of fires throughout the night to ensure the lions and elephants in the area stayed away from our bivouac site. During the night I could hear lions grunting and calling out from what must have been less than a hundred-meter distance. Maybe much less.

Zimbabwe SAS and 1st Parachute Group, Harare

After we finished our training at the Kariba Tracking Combat Unit, we returned to the installation on the airbase outside the Harare International Airport and conducted advanced airborne training for members of the SAS Light Infantry and the 1st Parachute Group. They were getting training in the US military's steerable static-line parachute system. As we provided the training with these units, we were also conducting training with the Zim PJI members in Military Free Fall operations.

Towards the end of our deployment, the Zimbabwean military held an event celebrating our joint training with a semi-formal meal in the Officer's Mess on the airbase. We also were given a briefing on one of the battles fought during the Rhodesian Bush War. The briefing was given by a former commander of the 1st Parachute Group (white guy) and its current commander (black guy).

What made the briefing unique was the fact that the former commander had been in command as the unit, under its previous Rhodesian status, had fought the battle against the current commander whilst he was leading one of the African nationalist groups during the Bush War. They seemed quite friendly now and struck me as odd to listen to them joke about having tried to kill each other during the battle.

I remember that one of my Detachment members asked the former Rhodesian commander about how they would locate the guerilla camps they would parachute into during their attacks. He told him that they would know about where they were located from gathering intelligence from the area's locals or captured guerillas, and then they would just fly around the area. Before long, he said, they would start shooting at the planes and once the bullets started coming through the floor of the aircraft, they'd jump out of the planes and start the attack.

That was a unique way to mark the Drop Zone.

My Friend the Farmer, NW Harare, Chinhoyi Area

Not all Africans are black. One of my former teammates from 5th Group had met a woman who was originally a Rhodesian. A white Rhodesian, or sometimes called a Rhodie. Most Rhodies found themselves in a volatile, or at least, unpredictable situation once Zimbabwe gained independence and formed its new government. Many left the country and emigrated elsewhere, mostly to South Africa, but also to England and even the USA, among other places.

My friend and former Teammate married the woman he met and we all became good buddies. Most of her family had left Zimbabwe during or after the war, but her sister had married a former Rhodesian Defense Force soldier who remained in Zimbabwe as a farmer.

During my second deployment to Zimbabwe I was able to spend a weekend on his farm outside Banket, near Chinhoyi. He had a nice farm and an average house, at least by American standards for a farm owner's property. In Zimbabwe, I suppose it was seen as a much nicer house and property in comparison.

Within the confines of his property, there was an entire village of local people. His farm provided employment opportunities for most of the people living in the village and allowed them to have a decent life with education opportunities for their children and access to whatever else they might need.

I was taken to several farms in the area and introduced to their owners. I saw some very impressive operations that really provided a lot for the economy and local opportunities. One place that really seemed fascinating was a rose farm. There were greenhouses with what must have been thousands of rose bushes. These roses were groomed by workers and then as they were ready, would be harvested, processed, and flew to various places in Europe for sale within hours after being picked, packed, and shipped.

It was really quite impressive to see the ingenuity and management that was required to make these various farm businesses run so smoothly. All of this contributed to the well-being of the entire nation of Zimbabwe. It was quite sad to see what happened once the policies of Mugabe started to destroy what had become a great example of how an African nation became so successful.

The former soldier and current farmer I met was also an exceptional man. Quite laid back and good natured, although he had been through a lot during the war. He opened up a little about his experiences, but one event stood out. He talked about having been on a reconnaissance mission where he and another soldier had crawled into a spot to watch the enemy through binoculars.

Apparently, someone found them as well and while he was looking off in the distance, the enemy that had sneaked up behind them shot him in the butt cheek. He said that the biggest surprise he got wasn't at being discovered, or even being shot, but at how badly it hurt. It may not have been how I would have told the story, but I never forgot his tale of being shot.

Now, so many years later, he and his family have moved to the USA after having his farm stolen, or taken by the government and handed back in small chunks to the black Zimbabweans. The farm no longer produces anything and I can only assume that the villagers who lived on the property, have either moved on or had to find other ways to earn a living.

Smoke That Thunders, Victoria Falls

Over the years that I deployed on various missions to southern Africa, I had the opportunity to visit Victoria Falls three separate times. I once drove there from Gaborone, Botswana with a Teammate, over a four-day trip during a break while training the Botswanan Defense Force on newly acquired anti-tank weapons. Twice I flew there from Harare, with my Detachment and members of the Zimbabwean Defense Forces of the Army and Air Force.

Victoria Falls is called the largest waterfall in the world. There is a wider waterfall, and another that is taller, but the combined width and height of Victoria Falls is not equaled. When the Zambezi River, which flows over the falls, is at high levels, the spray from the over 1700-meter-wide wall of water can be seen from 30-miles away. Likewise, the constant roaring of the water rushing over the falls and crashing into the river channel 355 feet below can be heard for a long distance. Hence the more popular local name, Smoke that Thunders, or in one of the native languages, "Mosi-oa Tunya."

I have been able to see the falls at both a high-water level and when drought had almost stopped the flow, leaving only a few places where the water continued to spill over the edge. One side of the falls is located in Zimbabwe and the other sets in Zambia. Both are very popular tourist locations and provide an excellent opportunity for enjoying the visit.

Flying with the Zim Air Force made the trip quite unique. As we were in a small group flying in a CASA-212 that was marked as the Prime Minister's aircraft, we not only flew to the falls, we flew all around and got to enjoy the sights from both sides of the Zambezi River. We even flew at a low-level right down the river and over the falls as the water disappeared into the huge crack in the earth and continued downstream.

One of the things I always noticed when visiting the falls was the lack of safety barriers. Only out by one end of the falls where during high-water seasons, the rocks are constantly wet, and very slippery, is a 3-foot high wall built of carved stones. The wall was built right on the edge of the cliff that drops straight down to the river's surface beneath the falls. The falls are actually rushing down the opposite cliff wall across the gorge from the viewing area and without raingear or an umbrella, you will quickly get soaking wet by the spray. For the rest of the area, if you wanted, you could walk right to the edge and step off without having to cross either a wall or a fence. When I was there, I didn't even see a warning sign telling visitors to stay away from the edge. Liability laws are obviously not as stringent as in the USA. Or perhaps, common sense prevails.

Besides seeing the falls and enjoying a natural wonder not found elsewhere in the world, the amenities are very enjoyable. The simple act of getting coffee and a cookie, or biscuit as it is called in the former English colony was special. The service and atmosphere of the hotels and high-end tourist attractions, while possibly unwelcome by a few, made the customer feel as if he were suddenly transported back to a time when Africa was still a distant land of adventure. It was selfishly enjoyable to find that suddenly, you were being treated like a famous explorer, or a great hunter on safari. It made a nice memory as you enjoyed feeling important and the center of attention, even for a few minutes while being treated to attentive, personal service.

That being said, wherever I went, I never forgot that I was in a foreign land and I never had to look far to see a stark reminder that I was in fact, in what was once called the Dark Continent, the wild lands of Africa.

Poem Published by Local Magazine, Harare, Zimbabwe

After my latest mission, I was feeling artistic and penned a poem about our visit to Zimbabwe and decided to send it to a local news magazine. The "Parade" as it claimed on its cover page, was "Zimbabwe's most widely read news magazine with nearly two million readers" and it contained a section titled "Poetry Corner."

The magazine had published my poem and mailed a copy of the magazine to me all the way to Ft Bragg, NC. There it was for everyone to see, page 44 of the August 1993 edition, was my poem titled Mission Memories.

MISSION MEMORIES

Blue skies over Africa filled with 'chutes of green and gray,
Special Forces and Para Commandos training hard throughout the day.

Shona, Tomba, Mandiki, tribes of Zimbabwe House of Stone.

Men of Carolina, tribe of Fort Bragg, Home of the Airborne.

Makuti, follow spore, heat and rain, hills and valleys, follow spore,
every man feels the pain.

Sleep at night, keep one awake, outside the camp lions roar
and elephants, the ground can shake.

Soldiers think of family and friends, so far away,
part of their heart and mind asking, "why do I stay?"

The answer is not to be found, it is not that simple.
One hears the call and follows.

The rifle, an extension of my arm.
A soldier and a weapon. Both together, complete the one.

"Elephants on the Drop Zone" one Commando was heard to say.
Under parachute, he flew to tall trees, that day.

Para group, Hot DZ, Maroon beret on their head.
The sight of the Paras fill the enemy with fear and dread.

American Green Berets, known around the world,
known to Zimbabwe.

American warriors train with warriors of the House of Stone,
each learning, each teaching, strength comes as our nations grow closer.

Zimbabwe military, American military,
elements of a great team.

As we depart, we remain joined in spirit,
the spirit of the warrior.

10th Special Forces Group, SFODA-061

Upon my promotion to Master Sergeant while I was assigned to the MFFS, I immediately contacted the SF Branch assignments office and requested to be reassigned to a line unit once again. My SWC tour was supposed to be a 4-year assignment and I was only at the 18-month mark, but as a newly promoted E8, I wanted to be a Team Sergeant again.

I was told that I could pick the SF Group I wanted and the reassignment would happen quickly.

I contacted the 10th Special Forces Group's Command Sergeant Major (CSM) and talked about an assignment. He was anxious to get me aboard and said he had a MFF Detachment in his 2nd Battalion for me. I was probably one of the most experienced and highly trained MFF qualified Master Sergeants in the Special Forces community at the time, having spent almost 10-years on a MFF SFODA, and had just completed an assignment as a MFFS Instructor. I agreed that the position sounded great and let the SF Branch know. I was soon on my way to Ft Carson and 10th Special Forces Group.

When I arrived and got to 2nd BN, I was interviewed by the BN CSM. I was quickly surprised to find that rather than placing me on the MFF Detachment, he was sending me to the SFODA that had the additional duty as the Group's sniper committee. The SOTIC Detachment, or Special Operations Target Interdiction Committee (or Course), was apparently suffering from a lack of (his words) Adult Supervision. In other words, there was a mess in progress and he wanted me to step in and fix it.

I was not happy to hear this, at all. But, there was no alternate option, at least none I wanted to try. The SOTIC crew was a mix of SF soldiers, most of them good guys, who had been mishandled by a dishonest and ethically challenged senior NCO. The Dctachment was commanded by a Warrant Officer who was also a good guy, but failed at correcting some very obvious problems within his small command. This was not that rare of an issue with Warrant Officers because many of them went the Warrant route early in their SF career and never gained the required experience as a senior NCO prior to becoming command eligible.

Long story short, I quickly discovered some serious issues with property accountability and fired the former senior NCO. I readjusted the training priorities and brought the Detachment's focus back to Battle Task mission requirements. Once that had been accomplished, along with 100% equipment accountability and reorganizing the real property tracking methods, we started on the Group-mandated additional mission of sniper training once again.

During this time, I also requested that I attend the Level One SOTIC Course at Ft Bragg, NC. As the NCOIC of the Group's SOTIC Committee, I felt that I needed to be well-trained in the subject matter. I also received a second course slot so that I could bring my assistant Operations Sergeant with me and he could attend the course as well.

Once I returned to Ft Carson and began rebuilding the SOTIC training capabilities, my Team Leader was reassigned and I got a new Warrant Officer as a Team Commander. As fate, or some other element of chance smiled, I ended up with a Warrant who not only was exactly who I needed as a partner in the leadership division, he became a life-long, great friend.

Soon, the 10th Group SOTIC was producing sniper course graduates and training events that were on par with the Level One course in Ft Bragg at the Special Warfare Center and Schools. Plus, we conducted routine SFODA missions, to include deploying to Bosnia, for six-months on one mission.

Needless to say, I stayed busy. I even finished up my military career in this position as the acting Commander once the Warrant Officer retired. For over six-months, I played the part of an "officer" during Command and Staff meetings.

During my tenure with the 10th SFG(A), I was the assigned Team Sergeant, but for a while, I also held the position of the Team Commander. The Detachment was a Company level team, SFODA-061, but was also considered a Group asset, the Special Operations Target Interdiction Committee (SOTIC). At times, I answered to three bosses, the Commanders of the Company, the Battalion, and the Group.

My Team was responsible for training Group snipers, as well as select personnel and qualified snipers from other units across the spectrum of Army career fields, but primarily from the Special Operations Forces. My Detachment conducted training in basic sniper operations, producing a military-wide qualified sniper at graduation. We also conducted sniper sustainment train-ups, and advanced sniper operations. This in conjunction with the traditional operations for a standard SFODA.

An additional requirement was to conduct Research and Development of tactics, techniques, and procedures for sniper operations. R&D was also conducted by communicating with manufacturers and representatives from the shooting and hunting communities of the commercial world. I attended the annual US Shot Show where I was able to gain the latest information and insight on weapons systems and all related equipment and materials available to support the SOTIC mission.

Some of the leadership and multi-tasking skills I learned while serving in this position were greatly beneficial later on as a contractor. Many times, I had to balance expectations and efforts between company and customer bosses, or Military and State Department bosses. My time spent at 10th Special Forces Group was more beneficial than I thought it would ever be while I was experiencing it first-hand.

ECWT 97, Camp Hale, CO, SFODA-061

I was introduced to skiing in 1997 after I had arrived at the 10th Special Forces Group located at Ft. Carson, CO. I had spent the previous 15 years in other SOF units that focused primarily on desert operations.

As I arrived in my new unit I was given a Detachment, along with a couple of duffle bags stuffed full of cold weather equipment that I had never used before. And yes, that included skis and snowshoes. These items were quickly placed into the back of my locker and I got on with the current business at hand, soon forgetting about all the white-colored gear. But eventually, summer changed to winter and all at once I was told to wax my skis. My first ski trip turned out to be quite an adventure.

We planned for our mission, which was called “Extreme Cold Weather Training” or ECWT to be 10 fun filled days in Camp Hale, CO where the temperature in January never got above zero the whole time we were there. As we left the base camp and headed out for our mission site, my new set of skis and I got acquainted very quickly, along with the 70 pounds of equipment I was dragging behind me in a covered sled and the 50 pounds of weapons and equipment I had strapped to my body. The first two days were limited primarily to flat terrain and gentle slopes as we traveled the 10 miles to get into our target area. After that we started going up. I now know how a turtle feels when he has been flipped on his back, because that’s what I experienced every time I came off my feet and ended up buried in the deep powder as my legs and skis remained above me in the light of day, or night as it was a few times. As we practiced ambushes and raids while wearing skis and snowshoes, I soon realized that fighting a war in the snow would only make a bad thing even worse.

During our trek through the area, we built a snow cave which could be used as an emergency shelter in certain situations. We began by building a huge pile of snow and then digging into it to hollow out a shelter. It would have worked great on a smaller scale, but we had built a grand hotel. Eventually as it kept being built higher and wider, we reached the limits that snow-alone could handle. There were two teammates inside the shelter when with a big whoomph, it collapsed. The guy near the doorway was blown outside like he'd been in a cannon's barrel. The other poor soul, lay buried under a ton or so of snow. We quickly and furiously started to dig him out and had him free within a couple of minutes. It was a good experience, at least from the aspect of finding out why you don't want to build too big a snow cave.

For the nights, we would stop in a spot with as much concealment as we could find, and then dig out an area a couple of feet into the snow that was large enough to pitch our two-man tents inside. With all of our gear and a couple of sleeping bags, the tent was pretty full. Most everyone carried an old Nalgene water bottle to keep with each of our sleeping bags. It was a little tricky, but using the bottle for a night-time urinal was a lot better than dressing and trying to get out and away from the tent to do your business.

Besides the pee bottle, you also kept your drinking water and boots inside the bag to keep them from freezing overnight. The moisture from your breath would have the inside top of the tent covered with ice when you got up in the morning. Also, you didn't want to put on frozen boots, on top of everything else, when you woke up.

Working in that type of environment kept everyone busy. If you weren't training, you were melting snow to drink, heating food to eat, or making sure you and your gear, and everyone else's, was still functional and accounted for. The weather was as much an enemy as any human could be when the cold and conditions became so extreme.

After a few days I had started to "master" the basics of surviving and even functioning at a working level fairly well. Movement, with skis or snowshoes, became easier and started to feel more natural. So far, all of our major movements had been pointed more-or-less, up the mountains. It may sound odd, but for me at least, it was easier skiing up the mountain than it turned out to be skiing back down.

I was lucky in that I had three actual ski instructors on my team. Working with them, I was able to quickly get prepared to start back down the mountains so that we could eventually make our way back to the Company HQ's base camp for exfiltration and EndEx, or the end of the exercise.

As we started down one of our return routes, I was told by my teammate instructor that we were skiing off-pista, and that our routes would generally be considered double black-diamond slopes. Thankfully at the time I didn't realize just how challenging or even dangerous that type of skiing could be. Between the trees, the boulders, the cliffs, and the occasional avalanche area, we finally got down to relatively flat ground within one of the high valleys again. We now had only another 10 miles or so to ski in order to get back to our exfiltration point. We had been in some very demanding and hostile conditions for almost nine days straight. I felt like we had accomplished everything we set out to do, and no one had gotten hurt, had any cold weather injuries, or lost any equipment. We had all the earmarks of a successful mission under our belts and I was thinking that it would be great if only we could finally catch a break. That's when we heard the roar of a snow mobile, or as the Army called it, a snow machine.

Not knowing who might be headed towards our position, we quickly scrambled out of sight. As the snowmobilers got closer, we could tell it was one of our own snow machine teams headed back to base camp. We stopped them when they got near our location and coordinated for assistance. It wasn't a ride, but it was the next best thing, or so I was assured.

We would skijor back to camp. This was another first for me and while it sounded good in theory, I quickly found out just how hard it was in practice. We loaded all our sleds behind one of the snow machines and then tied long ropes behind two more. We split up and put 5 men on each of the ropes. Again, I was the one with the least experience. Being the less-dependable skier, I was the first one on the rope line and closest to the snow machine. I was told that it was the preferred position and the easiest to handle as a beginner.

The theory was that I would have to deal less with the whip-type action associated with being nearer the end of the rope. I quickly found out that being pulled about 30 mph while hanging onto a rope, was not an easy task. I was more worried about falling and getting smashed by the four guys behind me than I would have been worried about flying off the end of the rope. It was a quick, but exhausting trip back to our Company's base station. As we pulled into the camp and rounded the last corner, in front of all the support troops and other Detachments that had already gotten back, I finally did slip and took out half my own Detachment and part of the base station's command element that had come down to watch us maneuver behind the snow machines.

I can truthfully say that this entire process, an immersion into the world of ice and snow, was one of the fastest ways to learn to ski, but I wouldn't call it the best way, nor recommend it for just anyone. Going up the mountain on a chairlift and then skiing back down on a groomed slope with nice equipment and no extra weight in the form of a rucksack or weapon is much closer to any idea of fun. Especially when it is followed by a nice warm meal in the ski lodge and a few hot toddies in front of a roaring fire in the lobby.

Due to mission requirements and some real-world situations, I only attended ECWT once, and never was able to participate with the unit during any of the actual ski train-ups during my time in 10th SFG. At least once a year, if not more, the unit would deploy to one of the ski resorts in CO and the personnel would actually be instructed on various levels of skiing techniques.

Looking back, as a CO resident who doesn't ski now that I'm retired, I wish I'd have been able to take advantage of the opportunities. However, I did get some great memories and extraordinary experiences at Camp Hale that winter.

Counter-Sniper Course 02-97, 1st Infantry Division, Grafenwöhr, Germany

This was a very unexpected requirement or tasking that came down the chain of command from echelons above God. I was notified about four days in advance to prepare for this event. I had the option to refuse and have it passed to another organization, but this was an opportunity too good to pass up. I quickly summoned my Assistant Operations NCO and told him we needed to be ready to fly to Germany in three days to present this one-day course. In his typical fashion he just smiled and said "no problem" as he started gathering notes and slides for a presentation.

The target audience was to be the entire leadership element for Company level Commanders and above for the 1st Infantry Division, the Big Red One. The course, which really was to be a class, would be given by me and my Teammate to the audience from the stage of Graf Theater located on Camp Aachen. We flew into Munich and drove to Graf, staying there for Friday and Saturday nights. Friday we were able to go out for some great German food at a local Gasthaus and then make some last-minute adjustments for the class to be given the next day.

That Saturday morning, we hit the ground running and for the next six hours or so, we conducted our class almost entirely from memory while conducting an ad-lib, back and forth conversation between us and the members of the audience.

The biggest push-back we got was when we suggested that as a counter-sniper measure, the Command element should incorporate a no-salute rule anytime they were outside a secured location. They were deploying to Bosnia, not Iraq or Afghanistan as this was pre-9.11.2001. The idea of not saluting was a no-go right away from the Field Grade Officers.

However, the rest of the impromptu class went quite well and we were made to feel much appreciated as we wrapped the class up at the end of the day. We each were presented 1st ID coins by the Division Commander and then we congratulated ourselves with another trip to enjoy some German beer and food. The next morning, we drove back to Munich and flew back to Ft Carson in time to get a little sleep before starting a Sniper Sustainment Course for our Group snipers. That was quite a busy weekend, but it was a good trip and well worth the effort.

Operation Joint Guard, Bosnia-Herzegovina, SFODA-061

From March until September, 1997, I was deployed with my Detachment and living with the natives of Bosnia-Herzegovina as we conducted Joint Commission Observer (JCO) duties throughout our sectors that were controlled by various former-warring factions. Our primary duty was to be an official conduit for direct communications between the Commanding General of the Multi-National Division – North (MND-N) of the Stabilization Force (SFOR) and the local military, civilian, militia, and other governmental or non-governmental organizations within our assigned sectors.

Our secondary mission, and somewhat "unofficial" was to collect information and Intelligence, also known as Ground Truth. This intelligence collection effort resulted in our submitted reports often going directly to the JCS or even the White House in their raw form. This type of activity, no matter how low-level or covertly conducted, did not go without notice, some of that being unwanted and even dangerous.

My JCO Team was comprised of myself, my Team Leader, and four other NCOs from my Detachment. About half-way through the deployment, a couple of my Team members were rotated back to HQ to assist the Command Element and I got another of my regular Detachment members, along with a US Navy E6 (Petty Officer) from SEAL Team 8. He was filling the slot for our communications NCO, as well as being a JCO team member.

My JCO Team changed sectors about one-third of the way through our deployment. We had started in a town in the west of B-H named Vares, a Bosniak or Muslim military-controlled town, surrounded by Croatian military elements inside the B-H Federation. The Croatians were Catholic and during the active fighting, they would often switch allegiances, sometimes fighting against the Muslims, sometimes aiding them as they both fought against the Serbs. The area around Vares was calm enough that after a couple of months, my JCO Team was transferred to a town named Vlasenica, located within the Republika Srpska (RS), under the control of the Serb military.

This area within the RS was a bit of a hot spot and had not had a JCO Team located within the sector prior to our arrival. I had to arrange for interviews with new Serb interpreters, as well as locate a place to live. As luck would have it, there was an International Police Task Force (IPTF) station also located in Vlasenica, and one of their interpreters was able to assist us in both, locating our new interpreters and finding a suitable house in which to establish our base of operations.

There Was Resistance, JCO Vlasenica, RS

Vlasenica was a hub of the local region in various ways. Physically, it contained a major intersection of the two primary highways that provided north-south and east-west access for the residents. Likewise, it was providing the same access to the SFOR convoys and patrols, most of which originated from a few miles west and just across the Zone of Separation (ZOS) between the RS and the BiH Federation. This SFOR base housed the US element that patrolled the area on both sides of the ZOS. My Team Leader and I would regularly attend the unit's security and Intelligence briefings as we maintained a close relationship with the Command element. This was also the unit that was tasked to provide our quick reaction force (QRF), should we need one. There was also a small PX and medical dispensary we could use, as well as a mess hall, should we want to eat during our visits. This was also our fuel point and as a result of the miles and miles of patrols our JCO Team conducted, our two civilian SUV-type vehicles required refueling about three times a week.

The profile of the JCO Team was much different when compared to the normal SFOR patrol. While the SFOR patrol leaders would conduct meetings with many of the same authorities that our JCO Team would meet with, the type of information, and more importantly, the atmosphere of the meetings themselves were not the same.

The SFOR patrol would roll into town with an armored convoy, protected by mounted machine guns and various other displays of military strength. Likewise, the members of the patrol would be wearing all of the regular "Battle Rattle" of armor vests, rifles, and helmets. If there was any reason to expect tension or push-back, there would often be helicopters circling overhead as well. There was usually not a lot of two-way discussions between the SFOR leaders and the local leaders.

The JCO Team, on the other hand, was the potential ally of the local leaders when they had a grievance or complaint they wanted to pass along to the SFOR Command element and we were not viewed nearly as much as an adversary as were the regular SFOR soldiers. While we wore US Army uniforms, we didn't wear any rank insignia, didn't wear headgear (hats), and we didn't carry any rifles or wear body armor. We did carry our pistols, but they were concealed beneath our uniform top. My primary JCO patrol consisted of me and my Team Leader (TL), along with one interpreter. My TL and I were great partners as JCO Teammates. The TL much preferred to attend the meetings in what might be called the secondary position. He would take notes, watch body language, and help control the flow as I filled the role of the primary speaker, or at times, interrogator. In spite of what some would consider a role-reversal between us, we never lost sight of who the TL actually was and what our roles on the JCO Team required. As such, we would schedule the meetings of primary importance for our attention during our daily patrols.

The other four NCOs rotated between patrols and house security, which included radio monitoring and report preparation, along with getting a chance to relax a bit. The meetings tended to be quite stressful and for the younger, more junior personnel, it was not a mission set that most enjoyed. The TL and I, however, had both completed extensive training and had experience in these types of low-level source, or Ground Truth collection operations, as well as planning and conducting meetings of this type with non-traditional military and civilian members. But not everyone was a fan.

While assigned to the Vares region, we actually brought the leaders of the local Croat and Bosniak military units together. We even hosted a dinner at a local restaurant and had the former enemies toasting each other. In fact, I had brought a replica of a Native American pipe and some tobacco to Bosnia with me, just in case an opportunity presented itself. I brought the pipe out at the dinner and passed it around the table as we all took a couple of puffs and proclaimed a newly agreed to, push to maintain peace. In Vlasenica however, this was not a majority opinion.

Soon after we set up shop in our new digs, which was a single house that had held two different tenants before, one family on the upper level and one on the ground level, we received incoming small arms fire. The shots had come from quite a distance and hit the house without causing any harm. We immediately installed some anti-grenade and small arms barriers over the large windows. We placed heavy curtains over the others and adjusted some of our in-house security postures.

During one evening, the SEAL attached to our JCO Team had stepped outside and was attacked by an intruder who had been hiding in the yard surrounding our house. He was stabbed in the shoulder by an assailant swinging a farming scythe that had been left hanging in a tree by the former residents. As he fell, the SEAL pulled his pistol and fired three times at the attacker. I was in the upstairs area of the house and ran down to where I had heard the gunshots.

The shots had missed the attacker as he ran away, but had hit the neighbor's garage door. As I got to the spot, I found our medic had already started to assist and was bringing our injured teammate into the house. While the wound was deep, it had missed all vital spots and he was able to return to duty after a few hours at the SFOR dispensary.

The next day, we received our requested security package and installed some sensors and remote cameras around our house. Additionally, we were offered a squad of Infantrymen from the SFOR base to come and spend the night on guard duty. We accepted the assistance and for the next couple of weeks we had some of America's finest for guests each evening until after daylight. They greatly enjoyed the opportunity to not only be away from the base camp, but to be away from the regular Army mentality, and I am positive that we influenced a couple of them to eventually volunteer for Special Forces later on. The same day, our neighbor whose garage had been shot decided to leave and go live with relatives. I can't blame him, but we were not winning many friends on our street.

Things settled down and went back to what we considered normal. But it didn't last long.

Once more as it grew dark, someone drove by the front of our house and threw a firebomb, a Molotov Cocktail, onto the hood of one of our vehicles parked in the driveway. One of our teammates watching over the area saw the vehicle stop, but didn't fire at it as it took off. Apparently, he determined that it looked like it could have been teenagers and that there was at least one female heard yelling and laughing. It seemed like there was a good chance that the act was a bad prank, rather than an attack by a more serious threat. Still, concerned that there could be a sniper waiting for someone to run out of the house to fight the fire, I kept everyone inside and under cover as we established a security perimeter in the event of additional attacks.

Our neighbor on the other side of our house decided to help out and came outside with his garden hose to fight the fire which was quickly engulfing our vehicle. He was yelling for us to come and help, and then became angry because we wouldn't help. Soon he realized that his garden hose was not going to make any difference and he went back into his yard. The fire department didn't come and soon the fire burnt itself out. Thankfully, the fire didn't spread to our second vehicle and only melted a little bit of trim.

As soon as it got daylight, the SFOR base sent a tactical wrecker out, along with a security patrol, and picked the burnt hulk up and carried it away. By the next day, we had replaced the burnt vehicle with another of the same make, model, and color. As it sat in the driveway we watched for someone to drive by and do a double-take, but we never found out who had thrown the fire bomb.

Mafia, Law Breakers, and Local Corruption

Besides rubbing a few local villagers the wrong way, we were also starting to upset some of the more serious criminal enterprises in the area.

While we didn't directly do any type of anti-crime operations, the fact that we were conducting operations that involved many of the local leaders, official and otherwise, we had impacted some of the organized crime activities, at least according to some sources with the IPTF and the local RS police force.

This element of the operation didn't really impact our operations and all we did with the information was to raise our levels of personal and mission security. It just became one more possible point of contention to consider as we conducted our daily rounds as we patrolled throughout our area of operations.

Srebrenica and the Muslim Massacre

Srebrenica was a town in a mountain valley to the east of Vlasenica that was infamous for a massacre of its male Muslim population by the Serb army during the active conflict in BiH. It still contained a majority of Muslims and was the proverbial raw sore in an otherwise healing countryside. On occasion, an SFOR convoy would patrol through the town, but it basically was left alone and on its own. Srebrenica became a major focus for my JCO activities.

The first few visits into Srebrenica and the neighbor towns or villages were met with outward hostility and skepticism. However, we, my TL and I, continued to return to the city hall and police station to discuss the situation and possible actions that could be taken to improve life in general. Soon, we were parking our vehicle near the center of town and walking a loop through the streets and stopping at stores, coffee shops, and cafes to discuss concerns, or at least listen to complaints from the townspeople who ventured an opinion.

Maintaining our low profile in both attitude and dress, we were easier to approach, but also made an easier target for the occasional angry interaction. There was still a lot of anger and grief inside of Srebrenica over the deaths of between 7000 to 8000 Muslim men and boys in July 1995, only two years prior to our arrival in the town. The Bosnia Serb Army, under the command of one of the primary targeted war criminals Ratko Mladic, had forced these people out of town and then killed them in an attempt to rid the area of all Muslims.

During our visits with both the authorities and the average people still living in the town, we eventually found our way to some rural sites that we determined were possible mass burial sites. Locating these sites allowed for attempts to identify the deceased and to provide the remains with proper burials.

A Car Crash and Hospital Visit, Milici, RS

One morning while out on patrol, we ended up behind an SFOR convoy headed south through Vlasenica towards the town of Milici. We were listening to the convoy's radio traffic as we followed them along the winding road, hoping that they would soon pull off or turn onto one of the side roads so we could pass and get going. It quickly became obvious that the convoy commander was one of the young Lieutenants we had met from the SFOR base just across the ZOS.

As we continued to follow, we heard someone in the convoy radio to a vehicle following them that it was clear to pass. Apparently, a civilian vehicle had gotten into the convoy and now the ones in the back were attempting to pass it to maintain convoy integrity. I told my TL, who was driving as usual, that this was a bad idea and probably wouldn't end well.

Just about then we heard someone say that it was clear to pass after the red car. Someone apparently did not hear the last part of the transmission concerning the red car. Seconds later, an up-armored Humvee with a .50 caliber machine gun in the turret, had pulled out to pass a civilian car on a blind curve and crashed head-on into a small red car being driven in the opposite direction by a local resident. The Humvee was clearly the winner in that match-up.

The convoy immediately stopped and we drove around the vehicles, sounding our horn as we moved past, on the way towards the front of the convoy as quickly as we could. The Convoy Commander, the LT we had just been talking about, was pretty close to freaking out. None of his people were hurt, but the man and young woman in the car were seriously injured. We parked our vehicle to block the road and jumped out to assess the situation. I still can clearly see what we had before us, as if I'd studied the setting for hours instead of the few seconds that passed as we began to react.

The small red sedan was angled across the lane of the highway in which it had been traveling. The front bumper and fenders, as was the hood, all smashed back against the firewall of the engine compartment and a large spiderweb looking series of cracks covered the passenger's side of the windshield. Cucumbers and other vegetables were scattered over the floor of the car between the front seats and across the back. Dirt was lying across the blacktopped roadway from where it had come loose during the impact. The man was skinny and had short dark hair. He was probably in his 40s. The girl was young and quite pretty with long blond hair and a light blue dress or skirt. The driver's door was bent and still closed. There was very little noise and no smoke coming from the car. It was very sunny and bright and the temperature was perfect for a summer day.

We had quickly taken the lead role in the event and took charge of the situation. I had the gunner post road guards as my TL and I began to examine the people in the smashed car. The driver had crushed his chest against the steering wheel and was suffering from what appeared to be a collapsed lung or pneumothorax. His leg was also trapped between the seat and the crushed in front end of the car.

The girl, who appeared to be in her late teens, was unconscious and bleeding from a large cut to her forehead. As we worked to get them out of the vehicle the LT was trying to call in a MedEvac helicopter. With some help from the guys in the convoy, we were able to free the injured man's leg. My TL was checking him for further injuries as I removed the girl from the car and quickly assessed her for additional injuries. The interpreter was helping me get a bandage around her forehead when we decided the best thing to do would be to load them into our vehicle and drive them to the local hospital that was less than a couple of miles away from where the crash had occurred.

The town of Milici, just minutes away from where we were located, had a small, but well-equipped medical center. We had visited the facility during some of our earlier patrols. We carefully loaded the two injured people into our SUV and took them to the small hospital. Both of the injured recovered and we were able to assist them and the hospital administration make required contacts for addressing the damage and injuries and other associated costs with the SFOR personnel.

This entire event, becoming an impromptu ambulance and medical team, and then following it up by assisting in the red tape, helped us create a friendly contact in the person of the hospital director. He turned out to be a former military doctor who had been in the Bosnia Serb Army. He was also a current high-level figure within the recently formed RS political strata. This helped open doors for future meetings and information gathering and most likely saved at least one life, and helped prevent more anger towards the SFOR effort within the region.

During our follow-up meetings, we learned that the man had been at the market and was driving home with his groceries. The girl was a local resident of the town of Milici and had been hitching a ride elsewhere, this being how she was in the car with the man driving. It turned out that the girl was also an aspiring model, and because of the scar on her face, left from the crash, she was attempting to get plastic surgery and a good monetary settlement from the SFOR coffers. We were able to connect both of the people who had been injured with the appropriate SFOR representatives and after that we never heard what, if anything happened later.

A Trip to Sarajevo

Sarajevo was a different city before the Bosnian War. The Winter Olympics had even been hosted in Sarajevo in 1984 when Bosnia was still Yugoslavia. Sarajevo was a city where the Serbian, Bosniak, and Croatian ethnicity populations lived together in harmony. As it turned out, the harmony was a forced situation, instituted by the communist government of the Yugoslavian federation. There are three primary factions in Bosnia – Herzegovina (BiH). Bosnian Serbs (Orthodox Christians), Bosniaks (Muslims), and Croats (Catholics).

As the politics of the region and of BiH were changed via the voting booth and legislation, war broke out in early 1992. The Bosnian Serbs wanted BiH to remain a part of the former Yugoslavian federation while the Bosniaks and Croats primarily wanted BiH to be an independent nation. Quickly, the people of Sarajevo and of the entire country of BiH split into religious-based factions and the Bosnian War began.

During the initial start of violence, Sarajevo suffered a division of the people and the Serbians departed the city, either voluntarily or they were forced out by the violence. Soon Sarajevo was primarily populated by Bosniaks and Croats. Also soon, the Bosnian Serbs laid siege to the city in an attempt to gain control. From April 1992 through February 1996, Sarajevo was cut off from the world by a ring of soldiers from the Bosnia Serb Army.

In Vlasenica, my JCO Team interpreters had both formerly lived in Sarajevo with their families. In fact, one of my interpreters, now separated from her former husband due to the war, had been married to a Muslim or Bosniak while living in Sarajevo.

Further, during one of the meetings I made with the local BSA military commander, we found out that he had also lived in Sarajevo prior to the hostilities breaking out. He and his family had to leave so quickly that they were unable to pack any of their normal household belongings and had left all behind, taking basically just what they could quickly throw in their vehicle.

My TL and I came up with an idea of how we might be able to not only build a closer relationship with the interpreters we had to trust in so many potentially dangerous situations, but also to build rapport with the BSA Commander in our town.

Over a period that included several trips into Sarajevo, we brought our interpreters back to see their old neighborhoods and to see how the city had been changed by not only the bombings, but by the new population majorities. It was both a pleasant and distressful experience for the interpreters. Both had grown up in the city and the war had created untold hardships. While we made the visits to Sarajevo, we also located the former residence of the BSA Commander. We decided that we'd make a visit to the new tenets of his former home.

He had lived in a medium-sized apartment on the 3rd floor of a large apartment complex nearby the airport on the edge of Sarajevo. Several of the apartment buildings had been destroyed during the war, but not the one in which he had lived. We brought our interpreter with us and made sure that she understood that she was not to bring up the fact that she was a Serb.

When we knocked on the door to the apartment, we were greeted by a middle-aged Bosniak woman. We told her that we knew the former occupant of the apartment and just wanted to see what the place was like. She immediately stated that she also knew him and that their families had been friends. Her family had moved into this apartment after the BSA Commander's family departed since it was larger than the one they had lived in. She asked if we would be seeing him again, and when we said yes, she presented us with a few bags of items to bring to him.

The bags contained items that couldn't have been replaced. There were wedding photos, pictures of their kids, diplomas and other official documents. She had kept all of the personal treasures that she could, in spite of the war between their people.

Once we had returned to Vlasenica and made an appointment to meet with the Commander. He was quite overwhelmed by our gifts, to say the least.

In fact, he couldn't say anything for a while as he looked through the photos and other items we had returned to him and his family. He didn't become our best friend, but our relationship became much better and his trust in our abilities to actually get things we said we would do, done, was raised by several levels. It was not long before we would get more information from him as we continued to work JCO duties in our region.

A Raid Across the Zone of Separation, Krivaja, RS

The next story, One Dark Night in Bosnia, is a story I began writing before we left the country and continued to add to it after my deployment. My Battalion Commander had asked the JCO Teams to write about an event that had happened during their operations that he could include in an After Action Report. I decided that I would write about an event resulting from a reported cross-border terror attack that I and my TL investigated. Because the story was written as a stand-alone story, there is information within it that has been duplicated in some of the other stories, or provides some additional information for subjects that have been addressed elsewhere in my BiH deployment memories. I believe you will find this next story quite interesting.

One Dark Night in Bosnia

This is a story about an event that took place during Operation Joint Guard in Bosnia-Herzegovina (B-H), while operating in the Republika Srpska, the Bosnian Serb controlled region of the country. The official mission our JCO Team was conducting in BiH as a part of SFOR, eventually came to an end in late 2004 when the European Union assumed control. The incident this story details occurred during the summer of 1997. As members of a Joint Commissioned Observer Team (JCO), a part of our mission was to provide Ground Truth. Ground Truth is the most current information available, originating from within an area and presented by the personnel operating within the area who have discovered and reported it. We reported to various SFOR command elements responsible for the diverse sectors throughout Bosnia-Herzegovina. Our primary customer was the Multi-National Division – North (MND-N), who at the time was commanded by Major General David L. Grange. He had just replaced MG Montgomery Meigs and both were US Army General Officers. Our team was responsible for the primary liaison mission and to act as the direct communications conduit between SFOR command and the local leadership, as well as monitoring the compliance of the former warring factions in accordance with the Dayton Peace Accords. In conducting our duties, we lived and worked in the operational area we patrolled.

Prolog

The Fourth of July was an interesting holiday that year in 1997. I spent it in a small town near Srebrenica, Republika Srpska, where the last major offensive of the Bosnian civil war was fought. Actually, the offensive was a major round up of male Bosniaks (Muslim Bosnians) who had been living in a proclaimed safe haven under the protection of the UNPROFOR (United Nations Protection Force) and eventually led to the massacre of approximately 8,000 Muslim men and boys which occurred from July 12 through July 18 of 1995. The massacre was carried out after the commander of the UNPROFOR, General Bertrand Janvier, refused to authorize the use of close air support to help defend the Safe Area, and eventually handed control of Srebrenica over to the Bosnian Serb Army General, Ratko Mladic. The massacre in and around Srebrenica, combined with the attack on Markale Market in Sarajevo on 28 August 1995, did lead to devastating air strikes directed towards the Bosnian Serb military. The resulting UN and NATO air strikes (Operation Deliberate Force) which practically paralyzed them and directly led to the Dayton Peace Agreement of 21 November 1995 were the reasons for most of the hard feelings our JCO Team encountered as we performed our duties throughout our sector, and specifically within the Srebrenica – Bratunac – Han Pijesak hot spots.

As we attended official meetings and mixed casually with the locals, we quickly discovered that the average Bosnian Serb didn't understand why an ally from WWII (meaning the US elements of the air strikes) would fight against them in their time of need, back during the Bosnian War. All this helped set the stage for the constant struggle our team had in dealing with our contacts while attempting to maintain an unbiased, observer's viewpoint in the accomplishment of our mission.

My team lived in a town named Vlasenica, in a two-story house near the center of the small city. We were just east, probably 200 meters or so, of the main road intersection providing access to all directions. Our operations room, interpreter break room, an office, two sleeping rooms (mine and the Team

Leader's) and a small latrine were on the upper floor of the two-apartment building. The ground floor consisted of three sleeping rooms, the kitchen and another latrine (bathroom) as well as some storage space.

Before we rented the house, it had been home for two families prior to the war. My six-man team needed all the space for our equipment and work requirements, not to mention Force Protection (security) concerns. Our JCO team consisted of my normal team organization, SFODA-061 of C Company, 2nd Battalion, 10th Special Forces Group (Airborne). Our Operational Detachment Alpha (SFODA or commonly referred to as an A-Team) was short a few personnel who were supporting our operations at the Company and Battalion Headquarters elements in Tuzla. In addition to the SF (Green Berets) on the team, we had one US Navy SEAL from Team 8 attached, giving us a "Joint Command" designation. I was the ranking Noncommissioned Officer (NCO) and therefore the Team's Operations Sergeant or "Team Daddy". We were commanded by Warrant Officer (CW2) Steve Hoffa who was not only the Team Leader but was one of my best friends. It was a good thing we were compatible because we spent many long hours together over six plus months and never really had what most people would consider a day off or a place to stand down and relax completely.

We had watched winter turn into summer and had been working seven days a week since we first got in-country. It was now July and we were attempting to take one day and make it as low-stress, as-normal-as-possible, even an enjoyable kind of day.

We kicked off the morning in this former Yugoslavian town of Vlasenica with its first American style 4th of July hamburger and hot dog cookout. We even invited several of the local VIP type of folk, with whom we normally interacted during official or scheduled visits. Most accepted our invitation for this event and came to the house. Perhaps they felt compelled to make a showing, but I think they were curious and simply wanted to see us in a more casual environment. One of our guests was the Vlasenica town mayor. A tall, thin man, "Slav" looked more like an undertaker than a mayor. He made me think of a guy who would get great enjoyment from pulling out someone's fingernails with a pair of pliers in some dank corner of a dimly lit basement. I wouldn't have been surprised to find out that he had done just that during the war. When he arrived, he was accompanied by the very nationalistic, local political party leader.

As we attempted small-talk, which is hard enough, but made worse when speaking through an interpreter. I tried primarily to discuss current issues and what was going on around the country as far as SFOR was concerned. The SFOR (Stabilization Force) was established as a NATO led organization, and was the successor of first, the UNPROFOR (United Nations Protection Forces), which was followed by NATO's IFOR (Implementation Force).

However, they both kept steering the conversation to the United States and how we had treated and continued to treat the Native Americans in the early periods of our history. Of course, they wanted to know how we would have dealt with the United Nations and an SFOR type of entity trying to force a mandate between the "white man" and the "red skins." This was an interesting conversation, but every statement needed to be carefully considered since there was no guarantee that our comments wouldn't end up in the news. It was real and it was fun, but it wasn't real fun. They may have been third-world politicos, but they were not behind any power curve when it came to intelligence and artful political spin. These guys knew more about how the American government operated and more about how to spin American history than the common American citizen did. The conversations were friendly, but pointed.

The director of the area's primary hospital, who was also the Chief of Surgery, stopped by for conversation and to have an American hamburger. We had met him after bringing a couple of local residents to his medical center in Milici, who had been injured during a traffic accident. The TL and I had watched a car get hit by an SFOR armored vehicle while rounding a corner and had jumped into the mix to assist.

Once we got to know him a little better during follow-up office calls, he began to show us pictures of when he was in the war. He had been a medical officer with the Serbs and was in photographs taken in the field and other places with some high-level figures like General Mladic, one of the war criminals on our list of things to watch for. He shared a lot of trusted information during some of our final visits with him. He even told us that he was also a Senator in the newly formed Republika Srpska (RS) legislative assembly, among other things. Meanwhile, back at the cook-out, the Doctor and his companion were more affable than the mayor and his political associate had been. The DR was more business-oriented and primarily talked about troubles they were having at the hospital with funding projects and collecting payment from NATO/SFOR or other governmental organizations for a variety of reasons.

Now, I'd never had an American Senator from the Federal or State level, or even a small-town mayor at one of my backyard cookouts in the US. It was a different and somewhat fretful experience knowing that every word I spoke would be taken seriously. We had our own interpreters at the cook-out with us and kept rotating them from "working" to relaxing with the remainder of the team and guests. There were also the locals who would try to speak English as much as we tried to speak Serbo-Croatian, but this was mainly with the non-official conversations being held by the more familiar people we worked around, or knew through the other UN or NGO (Non-Governmental Organization) teams in the area. One thing you always had to be cognizant of was that you could never be sure just how much English was actually understood by the people who claimed to not speak the English language.

As the cook-out progressed and finally started to wind down, LTC Alex Basevic, commander of the local RS Army unit headquartered in Vlasenica, visited us. He was not only the local unit's commander, but he also was the liaison officer with the RS Army Corps HQ who was responsible for the entire area we patrolled. He was educated by Yugoslavia's military academy and was a smart man. Since we all were soldiers, we always had an easy time talking with each other, as long as it was an "easy" subject that we were discussing. Alex mentioned, sort of as a side bar comment, that he would like us to drop by his office later in the afternoon so he could talk about something a little more official. He seemed to want to keep the conversation we were having at our house unofficial, simple and friendly. With the one mention of the follow-on visit request, he didn't talk about it again. In retrospect, I believe this was a sort of test us to see if we really were willing to go out of our normal actions in order to help (or try to help) as much as we had told him we would do, during the first few visits we'd made at his office. We were and we did.

The Incident

Once at his office later on that day, LTC Basevic of the Bosnian Serb Army (BSA) told me and Steve that there had been a border crossing incident which involved Bosnian Muslim soldiers from the Army of Bosnia & Herzegovina (ABIH) who had conducted a terrorist raid on a local RS village and resulted in one of the ABIH soldiers being killed. He gave us a location and then watched to see what we would do. Steve assured Alex that we would go check it out and make a report to our headquarters. The location was about an hour's drive away and not only right on the edge of the Zone of Separation (ZOS) between the RS and ABIH territory, but right between our own sector and the sector of the adjacent JCO team which operated as we did to our south. It was late in the afternoon already and we had been hosting visitors all day at the cook-out, but we decided we would get what equipment we needed, inform the rest of the team of what had been said and then drive out to get a feel for what we had been told. It sounded fishy, but we needed to know what might be going on if it were true.

We arrived at the location following the directions Alex had given us, as well as the military map we had. After asking a few questions to the people we found in the area, we were directed to a large brown wooden house on top of a hill at the end of a ridgeline. This was the last house in a cleared area that shared a border with the "no man's land" of the ZOS. No SFOR convoys had patrolled this area in several weeks due to increased requirements in other areas. The village was actually only a few houses scattered among the hilltops that the dirt road we had followed traversed. As we stopped and got out of our dark green 4WD Nissan Terrano SUV, we were met by a man in his late 20s who seemed very suspicious of who we claimed to be and what we were actually doing. As I explained the reason for us being there, again through our own interpreter, Steve and I noticed some activity at the periphery of the terrain we could observe from our position. Without any spoken directives, we moved into slightly offset positions where we were able to watch the ground and any activity that might take place to the rear of each other. Our interpreter was a young woman who had emigrated with her parents from the Balkan region to the United States at some point during the war. She had since grown up and had asked for a job as an interpreter or translator for the Department of Defense making full use of her native language skills. At this point she was blissfully unaware of the possible compromising position Steve and I believed we might have just gotten into.

We heard several men speaking just over the ridge near the house and a few others were seen moving in our direction from the other houses, and also from near the wood line. The house closest to us had a military type FM antenna on the roof and I could see at least two men looking out the windows at us. In our minds, Steve and I were already contemplating where we needed to fire and how we would break contact and escape if things suddenly turned sour. Outwardly we continued to act like everything was fine while inside we were spinning up and getting ready for whatever might take place. Before anything started to really develop, the man we were speaking to had heard enough to understand why we were there and then yelled up towards the house to say we were there to help. Very quickly we were joined by several others who now wanted to tell their versions of what had happened in hopes that we may actually be able to do something to improve their current lot on life.

As our personal threat level indicators gradually returned to a more normal state as we listened to and observed the people around us. We understood that these people were country folk who were simply afraid and seemed to be very happy to treat us as an official part of the system. It was apparent that they hoped that we could restore a more normal and civilized environment for them to live in and keep them from having to worry about incidents like the one that had sent us to this place.

It seemed the attack had occurred about 2300 hours, eleven o'clock at night civilian time, by five men dressed in the Federation's (ABIH) style of camouflaged uniforms. Apparently, they came out of the wooded area of the ZOS, crossed the open area to the first house and pushed or kicked open the door. The house they picked happened to be occupied by three or four men who were living there without their families while they were cutting trees in this area of forest. As the first man from the assaulting party burst through the door, one of the men inside the house who happened to be sitting behind the door as it was kicked open, immediately attempted to kick it back into the closed position. As he pushed, the door slammed into the second man causing him to trip and fire a three to four round burst from his AK-47 rifle. One of the bullets struck the first man who entered the house, who was directly in front of the shooter. Within a few seconds after breaking into the house, the assault team was suddenly retreating and attempting to drag their wounded comrade along with them, back across the ZOS and towards the wood line. The shots had alerted the other people in the small village and as the assault team attempted to escape, several of the loggers from other houses began to run towards the commotion and give pursuit. The fleeing soldiers left the wounded man behind when they got to a fence line and could not get the man over the fence. They did however, (according to the people we spoke to) take the wounded man's rifle with them.

The villagers took the wounded man, who had been gut shot through and through from the back to the belly, to a nearby house and began to question him. The villagers had refused to provide any medical care unless he explained what he had been doing and what was going on, so he spilled his guts, figuratively, as well as literally due to the gunshot wound.

He was in fact very distraught at having been left behind by the other members of his team and called them cowards and traitors. He told his captors that he and his team had been ordered to come into this village and kill all the Serbs they could find. If they were successful, they were to regroup and rally near another village in the area. He had been carrying a hand grenade, an unidentified unit marking on his uniform and a small rucksack with some food and personal items in it. He also had the rifle, which had reportedly been taken away by his fleeing comrades. Steve and I were told that the "terrorist" died from his wounds about an hour and a half after he had been captured. When we asked if the local police or the UN's International Police Task Force (IPTF) had been notified, they said that they had called the local Serbian military commander (someone we had not yet met) on the radio (at the other end of the antenna we had seen at the brown building) and the next day the police had come out and collected the body and equipment, and questioned everyone that was involved. We wanted to look at the house where the shooting had taken place but it was beginning to get dark and thought it best to leave and return later. About that time, we were told by another man who had just joined us at our vehicle (where we were talking with the villagers) that the "Colonel", the same local Serb military commander contacted via the radio after the incident, wanted to talk to us before we continued to investigate this incident any further.

Steve and I had already decided it would be better to return to this area in the daylight and as we were so far into the country side, it would be prudent to get back into town. We decided that we would stop at the military compound and visit the local BSA commander on our way back to home station. We received instructions on where the compound was located and then departed the village.

The Meeting

When we arrived at the compound we were asked to park our vehicle outside the gate and walk the 300 meters or so into the wooded area that surrounded a large log cabin style house. Most likely, this had been a lodge or sports club prior to the war and had been "acquired" by the military at some point to use as a command center. This was the apparent headquarters building for Colonel Vidakovic and his staff. Vidakovic was an apparent war hero who had made his way up the ranks via a blood covered path. He also was apparently still having a hard time trying to deal with the political situation that now called for diplomacy and intelligence, rather than battlefield bravery and simply killing the enemy. We were escorted to the briefing room, a large rectangle shaped room with an adjoining room that apparently connected with the kitchen. In happier days, this room must have been a formal dining room. The table and chairs were still there, but now rifles were hanging on the wall alongside pictures of the Republika Srbska's self-appointed president, Radovan Karadzic, as well as General Ratko Mladic, the top two people on our list of war criminals.

The Colonel was a large man, not fat, but big. He was about six-foot, four-inches tall and 250 pounds. He was wearing a camouflage uniform and was wearing a holstered automatic pistol on his belt. His staff was also sitting in the room around three sides (two short, one long) of the rectangle shaped table and they all appeared to be armed with AK rifles, either lying on the table, or leaning nearby against the walls. His obvious 2IC (second-in-command) was a very quiet man, but as the night wore on I got the impression he could understand English, but would not speak anything except Serbian. There were also a couple of lower ranking officers who I felt were there more to make us feel outnumbered than anything else.

As part of his staff, the Colonel also had a female lieutenant. She seemed to be in her mid to late 20s and while she had the typical Serbian post-war bad teeth, she was very attractive otherwise. Her uniform was also a woodland patterned, camouflaged field uniform similar to the others, however, she also sported high heels and a halfway-unbuttoned shirt that allowed the ample contents of her pushup bra to advertise her womanliness. She obviously was not there to take notes but acted more like a waitress.

Sandy was not our normal interpreter. She was an American citizen who immigrated to the US from BiH. She was at our JCO 4th of July celebration and volunteered to go with us on this trip. She normally would have been at the SFOR base. Unlike our local Serb interpreters, she was a Muslim, although you wouldn't know it by her actions. However, practicing Muslim or not, we didn't want anyone in this place to think she was of a Muslim background. I could sense she was quite uncomfortable from the unwanted attention she seemed to be getting from the Bosnian Colonel. Still, she continued to act completely professional and carried on as we had instructed her to do without showing any obvious fear or contempt towards the Colonel. Senada, or Sandy as she preferred, was actually very frightened. She always had us call her Sandy because Senada was a fairly common Muslim name. As I mentioned, she didn't practice any of the customary traditions, but still considered herself a Muslim. However, Vidakovic, a Serb who hated Muslims and had fought to kill as many as he could during the war, had somehow sensed that Sandy might be a Muslim and told her so. She claimed to have a Montenegrin ethnicity and that she was not a Muslim.

As she translated what was being said, I told him that she was an American citizen and to talk only to me and Steve, not to her. I would often tell the person we met with that the interpreter was only a voice in a box, and to speak directly to either me or Steve. I also reminded the interpreters of this quite frequently.

Vidakovic had been drinking slivovitz, or 'slivo as it was commonly called. Slivovitz is a very strong, peach brandy; most of it was also locally produced. He was beginning to become quite intoxicated. Steve and I were seated across the table from the Colonel and we had our backs to the exterior wall of the dining room. The first hour of the meeting had turned into something close to a scene out of a training film on what happens to Prisoners of War during an interrogation. Steve and I took turns and would answer questions with questions and utilize different techniques of guiding the conversation while allowing him to believe he was still in charge. I was seated at the corner next to the 2IC, who was still saying nothing and only listening. I had made a sidebar comment or two to Steve when I noticed that the 2IC officer definitely appeared to understand at least a little English. This complicated matters slightly as Vidakovic continued to attack us with his verbal assault. He had accused us of wanting to arrest him for war crimes and also stated that we were CIA operatives. He was just getting warmed up.

As for the CIA accusation, we were not surprised to get hit with this one. A news story published in a Belgrade newspaper a couple of weeks earlier had claimed that we were not soldiers, but CIA officers and listed us by name in the paper's article. This occurred shortly after we had experienced a surprise visit by some actual CIA personnel at our safe house in Vlasenica. They said that they were performing a vulnerability assessment on our rented house as a reason for the visit, but it was actually more of a sight-seeing tour. The story that ran in the Belgrade paper, which proclaimed our JCO team to be CIA operatives, had seriously damaged our credibility among the people of the local villages. This in turn caused several incidents which threatened the security of everyone on my team over the next couple of months.

As Vidakovic continued, he stated that all the Americans like us (ones that had the same JCO mission) who worked within the RS would be killed if any Serb on The Hague's war criminal list was arrested. Steve told him that our job was not to arrest anyone, but to facilitate communications between people like him and the SFOR commanders. The Colonel seemed to ignore what we were telling him and went on to state that he could not protect his people because if he left his compound wearing his uniform, he believed we (or SFOR) would arrest him.

He had been drinking a lot by then and seemed to be getting a little reminiscent of the war when abruptly, Vidakovic made the first direct threat against our lives of the evening. He touched his pistol and said during the war he had personally killed many people and had no qualms about killing more. He said he would kill us if he wanted to, and that he could shoot us or have us beaten to death. Our interpreter was a little slow to translate this for us, but body language can be very loud in circumstances like this. He was convinced that he was on the secret indictment list from which SFOR units had been using to plan and conduct HVT (High Value Target) capture or kill missions.

He assured us that he would never be taken alive should we decide to come after him. Again, he told us that he could have us killed before we were able to get out of this room. Steve and I were reasonably sure that we were listening to a lot of frustration and anger, mixed with a liberal amount of alcohol, but we paid attention nonetheless. While we were fairly certain that there was not a high level of immediate danger, we were not willing to let matters escalate. I told Vidakovic that I knew that he would not kill us, and asked if he expected us to simply sit there quietly and listen to him threaten us? I reminded him that we were there to discuss the reported incident in the village and that we wanted to prevent this type of cross-border activity from taking place.

He was having a hard time seeing us as a conduit between him and the SFOR headquarters. He had convinced himself that we were some type of secret hit squad and he was beginning to act like a man trapped. Steve finally told him that as American soldiers we had the firepower of all of SFOR at our beck and call. Furthermore, Steve continued, all of our people knew where we were and what we were doing, and if something as foolish as a fight were to start, did he really think he would ever be able to leave this room alive? Now while we didn't normally carry rifles and wear battle-rattle gear, Steve and I always carried a concealed weapon. The others in the room had not seen a weapon but knew that they had to assume that we were armed. Everyone, with maybe the exception of Vidakovic, knew that if a gun battle were to break out, there would be no winners. We had threatened that immediate air power and gunships were on standby, but in reality, no one knew exactly where we were or when we were due back. Steve and I were gambling and understood the stakes were quite high this night.

We now began to receive a lecture on how the SFOR treated the senior RS Army officers at check points and that they would begin to kill soldiers, rather than to continue to take these humiliations. Vidakovic told us the people were ready to rise against SFOR and that every hollow log in the woods had a rifle in it. He again accused us of being in his country to arrest war heroes that the "pro-Muslim American government" had placed on a secret list. He had worked himself up into quite a frenzy and had told us that we (all the JCO teams) would be killed if anyone from the secret list was arrested.

It was getting very late and we seemed to be at a stalemate. We had all had at least couple shots of 'slivo and several cups of espresso. All of the ingredients for an explosion were present in and around the room. There was a very profound pause in the conversation while we stared at each other all around the table and contemplated what the next step was going to be. Physical violence, while not the best or first choice, definitely began to seem to be a possibility. Vidakovic was sweating intensely now; his 2IC looked like he was computing long division in his head and the other two staff officers in the room just looked scared.

Steve and I were more irritated at this point, irritated and tired, and yet, while it shouldn't have been a concern, I was thinking about the negative ramifications I would suffer from my own headquarters if this event escalated into a shooting, or even worse. The last thing I should have been worried about was surviving a physical confrontation, only to wonder if our careers would fare as well. It was still pretty much a zero-tolerance Army for causing problems, regardless of the circumstances.

Meanwhile, Sandy was shrinking in both voice and stature, but she was trying hard to remain focused. After a few more seconds, which seemed like a very long time, the Serb female officer, who was also acting as a waitress, brought in more coffee and as she placed the cups around the table, she was speaking in a lighthearted tone and smiling. It wasn't translated for us, but I'm sure it was something along the lines of a suggestion to stand down from red alert.

At this point, the conversation turned somewhat lighter and began to be much less hostile. The other staff officers in the room all seemed to physically loosen up and start breathing again. Up to that point it had been very tense. The Colonel suddenly wanted us to stay and drink with him and to have something to eat. We refused, saying we were expected to return to base. He then offered the company of what he called his "aide-de-camp" for the night, to which the woman mildly objected, but looked towards us smiling.

Perhaps this was what she had mentioned earlier and our interpreter chose not to translate. Regardless, for now everyone's fingers had figuratively backed off the triggers. Vidakovic had put away over 10 shots of 'slivo since we'd started this meeting. He began to focus his attention and comments affectionately towards Sandy as she politely, but firmly ignored his attention. She also indicated to me that we should leave ASAP.

Steve and I began to contemplate our exit strategy and after another thirty minutes or so, we announced that we were leaving, stood up, and started walking towards the door. The side of the table we were seated on forced us to walk around the table and across the room, past all the Serbs. Vidakovic again tried to convince us to stay, and then he stepped between us and the doorway. We were all smiles at this point, but the underlying tension was still there and I could feel it start to surface again as we were standing in the hallway wanting to push aside the drunken commander and cover the 300 meters that was still between us and our vehicle.

Since I had arrived in-country, I had been watching the various customs and peculiarities of the region. One in particular that is very different from our own social behavior, is the kissing rituals between men that takes place at various times. On an impulse I decided that this would be a good time to play this trump card and I figured we could take advantage of the momentary surprise.

As Vidakovic again began telling us why we had to stay and eat, drink, and be merry, I grabbed him by both shoulders, pulled his head down towards my own and kissed him on the forehead. As he stood back up, I twisted him out of the way and said good-bye. I led Steve and Sandy out of the house and onto the dark path between us and the gate. Sandy appeared to be ready to faint. She asked me in an astonished way, why had I done that and whether I was aware what I had implied by doing so. I said yes, that I knew what I was doing, to keep quiet and continue moving down the path. We could hear Vidakovic laughing in the house.

We told the guard to open the gate and quickly checked the vehicle for bombs or other surprises. Once we got in, Steve started driving towards our home station. Everyone was quiet for a few minutes as we relaxed. The feelings of normalcy returned quickly following the several hours of very elevated stress levels. I finally asked Sandy what she had started to say as we left the house and walked to the vehicle. I expected that she was going to make fun of me for acting like a "native" and kissing the Colonel on the forehead.

Answering my question, she laughed a little nervously and said that, yes, men do kiss on the cheeks to say good-bye, but I had done something else altogether. By kissing him on the forehead and then departing, I had treated him much like a naughty child. The kiss on the forehead, rather than each cheek, showed I had forgiven his bad behavior of the evening. By my unintentional act of superiority, and by doing so in front of the commander's own subordinates, she thought we would not get off the compound. I said that sometimes things just have a way of working out.

In retrospect I don't think either me or Steve could have come up with a better way to not only defuse this particular situation, but to leave a positive influence towards other SFOR soldiers. This guy will long remember his encounter with Americans that dark night and it should be a good memory.

As we continued along the highway I suddenly realized that I was starving and we should try to find a place to eat. Of course, there were no restaurants open after midnight in the Republika

Srpska countryside in 1997. I had to wait until we got home to raid the refrigerator.

Follow up

We continued the investigation over the course of the next few days and it appeared that this raid did in fact happen. The small group could not be tied to any Federation unit and could only be labeled as independent terrorists or criminals. SFOR increased their presence-patrols throughout the area of the incident and no further reports of these types of activities were brought to our attention. It seems we passed the test as far as the BSA commanding officer of the town we lived in was concerned, and we also regained most of the respect and trust of several prominent figures within the area that we seemed to have lost after the Belgrade newspaper CIA story ran.

That being said, due to continuing Force Protection concerns and other incidents, when my team rotated back to the United States in September 1997, Vlasenica didn't get another JCO team. The hot spots of the Srebrenica region were handled by JCO teams from adjoining areas. Some JCO Teams went into the mission with a predetermined notion that no trouble and no incidents equaled mission success. My Team did nothing but stir the pot to see what we could find the entire time we were deployed.

During the course of the mission, I received both accolades and warnings of being fired. But, that's how I always rolled and I was always going to be myself.

The areas of operations.

My JCO Team covered from Vares in the west, to Srebrenica in the east; Zvornik in the north, to Zepa in the south.

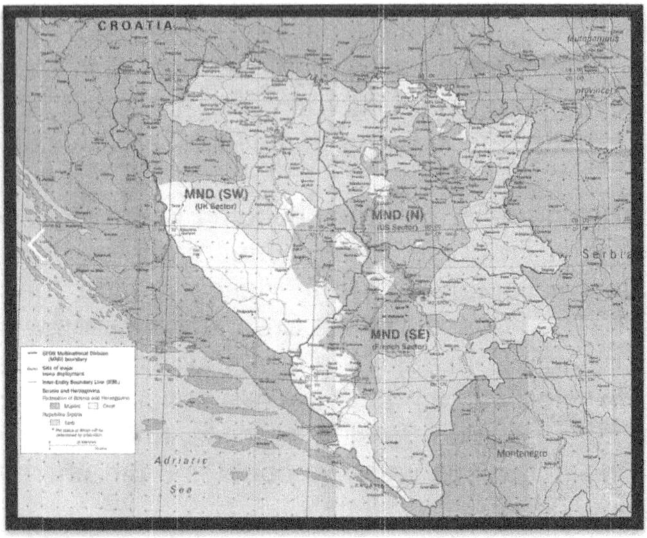

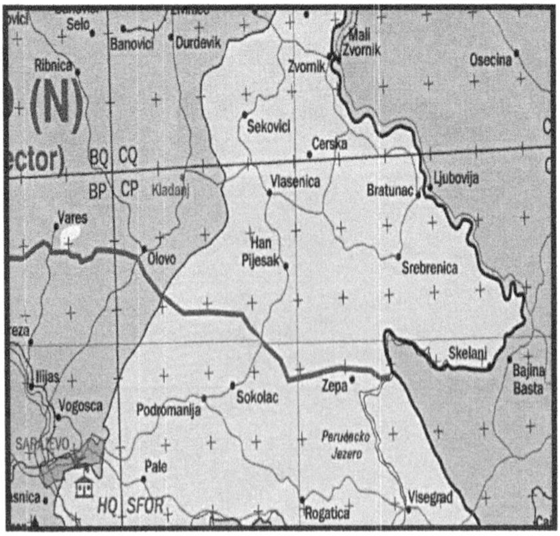

Another Poem

The following poem was written sometime after leaving BiH and returning to the USA.

While there, the personnel tended to start to identify with which ever element of the former warring factions they dealt with on a regular basis. My team was no different except that we spent considerable time working with all three of the factions. Being deployed first to the BiH Federation and then to the Republika Srpska, we came away with a more rounded opinion of the people and how they acted during the war, as well as how they acted and were treated after the war.

The poem Angels Crying shows my personal leanings toward the Bosnian Serbs, but none were without fault.

ANGELS CRYING

THE FLOWERS GROW SO TALL WHEN THEY FEED ON BLOOD THAT FLOWED FROM THE TORN BODIES OF THE CHILDREN OF THE REPUBLIKA SRBSKA.

GREEN GROW THE FIELDS WHERE HONOR GAVE WAY TO HOT STEEL AND COLD HATRED THAT WAS SPILLED UPON THE DEFENDER OF THE SERB TRADITIONS.

ANGRY AND CONFUSED, FEELING BETRAYED BY OLD ALLIES, DEFENDING THE RIGHT TO BE A CHRISTIAN NATION NOT OPPRESSED BY THE YOKE OF MOHAMMED, THE SERB DIED A THOUSAND DEATHS.

THROUGH STRENGTH, DETERMINATION AND A SPIRIT THAT IS AS THE TWO HEADED EAGLE, SEEING BOTH AHEAD AND BEHIND, THE FIGHT WAS BROUGHT TO THE EYES OF THE WORLD, AND THE WORLD TRIED TO HELP.

AS GIANTS CHARGED INTO THE ARENA TO RESTORE ORDER FOR THE INNOCENT AND UNPROTECTED, THE EAGLE WAS KNOCKED TO THE GROUND AND BRUISED, BUT NOT BROKEN, ONLY TO RISE AGAIN, WISER AND STRONGER.

WITH A CROWN OF KNOWLEDGE AND THE STRENGTH OF GOD, THE REPUBLIKA SRPSKA WILL OVERCOME THE OBSTACLES AND PREJUDICES THAT THREATEN HER, TO BECOME A FREE AND INDEPENDENT ENTITY.

BY USE OF THE MODERN SWORDS OF DIPLOMACY AND EDUCATION, DEMOCRACY AND CAPITALISM, THE PEOPLE AND THEIR TRUE LEADERS IN THEIR GOVERNMENT, WILL ENJOY FREEDOM AND LIFE FOR ALL.

WHEN THE TEARS FROM HEAVEN FINISH WASHING AWAY THE BLOOD OF MARTYRS, AND STRENGTH HAS BEEN ATTAINED WITHOUT VIOLENCE, THE CHILDREN CAN BE RAISED IN PEACE AND WITHOUT FEAR.

THE REPUBLIKA SRBSKA AND THE WORLD MUST NOT ALLOW THE CRIMES OF THE PAST, THE SLAUGHTER OF INNOCENTS AND ATTEMPTED GENOCIDE OF ANY PEOPLE OR CULTURE TO BE REPEATED. WE MUST PREVAIL FOR ALL OUR CHILDREN.

5th Annual International Military Sniper Competition, Ft Carson, CO, SFODA-061
The 10th Special Forces Group (Airborne) 04-11 December

1998

Oderint dum metuant, “Let them hate, provided that they fear”. This quote by a Roman poet, made over one hundred years before the birth of Christ can pretty much sum up what the military sniper has to believe about his chosen line of work. The mention of “sniper” in certain circles can be met with anything from coolness to outright physical violence.

There are different types of snipers, but the general public mainly considers them as cold-hearted murderers or cowards, afraid to face their enemy. A psychopath running uncontrolled by personal ethics or recognized rules of society. Nothing could be farther from the professional military sniper. A military sniper lives or dies by control and a deep commitment to God and country.

I have control, I have control, I have control…five, four, three, two, (BANG), one.

Execute, execute, execute.

This type of operation would be considered a good demonstration in precision planning, coordinating, communication, and without a doubt, shooting.

This is one scenario that a military sniper could find himself in. Another might be as a deadly voyeur, stalking and watching his target for days or weeks. His mission forcing him to become as close and intimate with his target, as the family and friends of his contemporary, who has been selected through detailed mission analysis, to be terminated. The military sniper, as opposed to the police sniper, will more often than not, find himself targeting an individual who is as ethically, morally and patriotically an equal to himself as his own teammates. The target, however, had the misfortune of being a member of an enemy organization.

Snipers employed in law enforcement, executive security and even certain special mission military units will always interdict persons who are known or suspected criminals. They can be proud and even boast of making the world a better place "one round at a time." The military sniper who patrols combat zones, whether in the rural or urban areas of the world only knows that the target must be interdicted in order to complete the mission. This responsibility truly takes a special kind of person, one who has been extensively trained and evaluated, both in his fighting skills and mental hardiness. These are the types of men who came to Fort Carson, Colorado to test their skills and share stories and ideas over the seven days of the 10th Special Forces Group's 5th Annual International Military Sniper Competition.

Thirty-four two-man teams, each of the 68 individuals were experts in long distance, cold bore, first-round-hit shooting. Each man had been trained to the highest level his respective branch of service or country was capable of doing. They had brought an assortment of weapons, but primarily the military version of the Remington 700 (.308 win or .300 win mag) was the standard. The competitors were limited to the government issued weapon from their parent unit, so there was no custom made, individually owned weapons used in the competition.

The contestants were primarily US Special Operations Forces with teams from all five Active Duty and both National Guard US Army Special Forces Groups, all three Battalions of the 75th Ranger Regiment, US Navy SEAL teams, Special Delivery Vehicle teams and a Naval Special Warfare Group. The conventional Army was represented by an Infantry battalion from the 4th Infantry Division as well as a Military Police Special Reaction Team. There were also teams from an Infantry battalion and Marksmanship Training units of the Army National Guard.

The NATO countries were represented by seven teams from Infantry and advanced training units from Canada and a team from a special mission unit in Norway.

The event got off to a smooth start with the exception of the Norwegian team, who had lost their baggage and part of one of their rifles during the flight to Colorado Springs. All turned out well when the baggage and missing rifle bolt showed up before the first scored event.

The first event was a known distance shoot that was fired from the 200, 300, 400 and 500-yard firing lines. This event was a National Rifle Association type event. The sniper teams participated as both a shooter and a spotter at each yard line. They fired from the setting, non-supported position from the 200 and 300-yard line and from the prone supported at the 400 and 500-yard line. The seated, non-supported position only allowed for the weapons sling to be used as support for the non-firing arm. This was one of the least liked events, as most snipers don't practice from this position during normal train-ups. The temperature during the event brought its own set of problems with it, as well. This competition is specifically run in the winter months to try and incorporate winter warfare training and techniques into the courses of fire. Most shooting matches are held in nice, warm weather, but the wind, snow and extreme temperatures associated with the areas of the world, not found on the equatorial plain, truly challenge a sniper.

This year's event was the first in a long time that was conducted in conditions creating the glare that snow and ice can cast into a rifle or spotter scope.

The second event was an unknown distance shoot on various size steel silhouettes from ranges of 250 to 1000 meters, as well as an urban scenario shoot with targets being paper cartoons that were composed of "terrorist" and "hostage" personnel. These targets were between 250 meters to 350 meters away from the varied shooting platforms and were located inside of windows of a façade. The range to each target had to be determined by the team within one minute prior to firing at it. No laser range finders were allowed and the snipers had to rely on mill formulas to size the targets and determine what dope (firing and aiming adjustments) to set on the rifle. The spotter had his work cut out for him as well, due to the

Colorado winter wind blowing across the range, whipping like a dragon's tail. Just because the wind was blowing at 10 knots, ¾ value left to right at the 300-meter target didn't mean it wasn't a full value right to left 15-knot wind at the 950-meter target. The unknown distance range is at the base of the eastern edge of the Rocky Mountains and the wind will change rapidly and often, even on a good day. This was supposed to be a competition to pit the best military shooters in the world against each other and Mother Nature was doing her part to ensure it was tough.

Wednesday brought about what is usually considered the hardest event, the team stalk.

The teams were required to stalk, undetected to within 300 meters of their pre-designated target.

The stalk lane had been planned and set up when the ground was shades of brown, but the snow that fell two days before the competition added a new dimension to the stalk event. The stalk lanes were down a slope that approached the "targets" as they viewed the area with binoculars.

Obviously, some lanes were better than others and it was up to the sniper teams to pick them.

They needed to conduct a good map recon of the area and then they could view the terrain from hiding spots, most of which were less than ten inches off the ground, while they walked, crawled or slithered to their self-imposed Final Firing Point. The snow backlit many of the sniper teams as they moved behind and through brush and bushes. Include a 25-40 knot wind and temperatures below zero, it was a very physically and mentally demanding three hours. Out of the 34 teams who started the event, about ten of them got caught by the sniper instructors on glass at the target sites. "Murphy" had showed up with snow and extreme cold. Add to that a poor choice of stalk lanes by some teams, or failing to stalk within 300 meters of the target and this event had effectively taken the hope away from 28 men of finishing in the top three places for the rest of the competition. However, it was never supposed to be easy.

As a new day dawned and these highly trained professionals had thawed themselves out and returned for another hard day of shooting, it was obvious that none of the teams who had gotten busted on the prior day's event were slacking off. Every team was as hungry for points remaining to be earned as they were the first day. That's the great part of this competition. It allows you to see up close, the superior mental stamina and moral courage these men of elite military units, from around the world, share. These are true warriors, as much as when the Knights of old jousted for honor and glory in-between campaigns and quests. The event being fired this day was named "Snaps and Movers." Again, from the 200, 300, 400 and 500-yard line, both members of the team would have to take turns at shooting and at spotting. At each yard line a target (head or partial torso) would appear for only brief seconds within a 10-meter front, as well as a 9 inch by 36 inch "walker" that would move left to right or visa-versa, across the same front at walking or patrol speeds. It is an awesome thing to see targets this size, and being displayed for only a very short time, be shot time and time again from farther away than the average person would notice someone standing and waving at them. In all fairness, there is technology that allows weapons systems to be much more accurate than was demonstrated by the snipers attending the competition, but the technology has its downsides as well. A sniper has the capability of providing the "user" with plausible deniability. A sniper can demoralize an entire unit and prevent them from accomplishing their prescribed mission, or at least slow it up. And then disappear like a ghost. A sniper can stay in place and report what he sees for an extended period, whether he is used in the fire-mode or not. There is a place on the modern battlefield for the sniper, now and in the future.

The final event was the highly secretive and anticipated stress shoot. All that was told to the competitors prior to the day of the shoot was expect it to be physical and emotionally stressful. Of course, this vagueness only added to the self-induced stress that teams were generating by trying to get early information. As the teams were randomly selected and brought to the start point with "anything you feel you'll need" to fire the course, they were instructed, in a broad sense, on what was required. The shoot began with a quarter mile sprint up and down a very steep hill. This was no small task in itself, considering Fort Carson is over 6000 feet above sea level and the air is mighty thin for a flatlander. From the first position, the team had one minute to identify, range and fire at the correct target, which was around 400 meters away. Only the first or second round counted for a score, and the time to run the event also counted for a portion of the score. If they took too long, they had to move to the next point with a zero for that particular target. The next point was several hundred meters through a ravine and up a rope climb, followed by climbing to the top of a four-story shooting tower. The tower was swaying several inches at the top due to the wind. The target was only a little over 200 meters away, but it was only a head that was visible, hidden behind a hostage with only about ¾'s of the "face" showing. Hit the target, get points. Hit the hostage, and you lose the same number of points. More stress. Upon climbing down the tower, the teams would have to run to the next point and from a prone supported position, fire at a target, again hidden among hostages, at about 650 to 700 meters away. Lastly, as the team rushed to the final point, they could look downrange and barely see the next target at 1000 meters away. The teams firing the .300 win mags had the advantage on this shot, but the .308 also reached out and touched it numerous times. The time ended after the final shot. Hit or miss. The scores were tallied up and the final standings roster was posted after we got back to main post.

After it was all said and done, the 5th place team was from C Company, 3rd Battalion, 172nd Infantry, US Army National Guard. The 4th place went to SEAL Delivery Vehicle (SDV) Team 2, US Navy. Third place was won by B Company, 5th Battalion, 19th Special Forces Group (ABN), US Army National Guard. The second place and runner-up sniper team came from the SEALs again, at Naval Special Warfare Group Two (NSWG-2), US Navy.

Finally, as with all competitions, the men who proved they had what it took to overcome any obstacle set before them and end up on the top of the pile, were evident. The First place team for the 5th Annual 10th Special Forces Group (Airborne) International Sniper Competition was from SEAL Team Two, US Navy.

My Team and I, the men of the 10th SFG(A) Special Operations Target Interdiction Committee, Operational Detachment Alpha 061, put many hours of planning, coordinating and back side support into the successful outcome of this competition. Many positive comments were made and much sage advice was given by all involved, and a good number from those who were not involved.

"Everything we do that does not kill us, makes us stronger." This was one of those events.

The planning and coordination that went into this mission was longer, more intense and further reaching than most deployments or training exercises. The satisfaction for a job well done was shared by all, as this was obviously the best competition of the previous competitions.

During the awards ceremony and dinner, one of the guest speakers ended up being a friend of the Special Forces community, one of the Denver Broncos. He was a nice guy, but just wasn't that impressive when he was compared to the real-life heroes of the Special Operations Forces and US Allies who were being honored. But, I suppose I'm a little prejudiced in that regard.

Running competitions and presenting sniper courses or training exercises wasn't all that my Detachment had to do. As an SFODA, the focus was on the O, or Operational. One of the requirements for 10th SFG Detachments was to be prepared to go to war in extremely cold or arctic climates. The training came with all the requirements to move in snow, so we did.

2nd BN, 10th SFG(A), Pre-Mission Training for Bosnia Deployment, 1998

Early in the Spring of 1998 my Detachment was tasked to plan and conduct the Battalion's pre-deployment training for another rotation to Bosnia-Herzegovina. The training would be conducted at Camp Guernsey, Wyoming.

My own Detachment had attended a similar iteration of training a couple of years earlier and we had several ideas on how to improve some of the more important tasks the Detachments would be required to perform once they were in country again.

During the earlier iteration as my team attended training, the priority had been on tasks such as small team or small unit tactics that focused on shooting rifles and even working with close air support. The actual mission on the ground once deployed, didn't really mirror the training we'd received, so having been tasked with the mission of preparing the Battalion to deploy once again, we were adjusting a few of the tactics, techniques, and procedures (TTPs).

Not to say that the training we had attended before wasn't good training, or that it wasn't fun training. It was quite the opposite in reality. It was always good to spend days on the range with minimal restrictions and maximum amounts of ammunition.

One of the more spectacular training events that I participated in was vehicle-based. A check point had been set up with targets representing a hostile force. Our practice sessions included driving a civilian SUV along a dirt road and, as we would be rounding a corner, come up on a hostile roadblock. The scenario required the driver to do a J-turn, which is an evasive maneuver to get turned around and headed back in the opposite direction very quickly. In our scenario, it was also conducted while under simulated weapons fire, requiring us to return fire at blockade.

We had been practicing this station for a while and we kept switching up the roles of driver and shooter. During this particular run, I was in the passenger seat and my Assistant Operations NCO was driving. I was armed, as I would be on a routine JCO patrol, with my M9, a military Beretta 9mm semi-automatic pistol. My teammate was driving a brand-new Ford Explorer, a rental, with less than a couple of hundred miles on it.

As we hit the point in the scenario where we encountered the road block and started receiving fire, he slammed on the brakes and started racing in reverse as I began to fire out the side window at the targets positioned around the road block. As my driver spun the wheel and swung the vehicle around 180 degrees to face in the opposite direction from which we had been travelling, a rear tire blew out. When the rear tire blew, the SUV launched into the air and very roughly landed on its roof in the middle of the roadway.

As this was happening, and thankful that I was wearing my seatbelt, I pulled my arm back inside the vehicle and secured the pistol, pointing it at the floor as we slammed to the ground.

The first thing that happened was that my body armor, on my now, upside-down torso, slammed into my chin and mouth, pushing one of my teeth through my lower lip. The roof had not held up very well, and was now smashed down, or rather up, against our helmeted heads as gravity tried hard to pull us out of our seats through the seatbelts.

The next thing that happened was my medic and a couple others were pulling on me in an attempt to get me out of the upside-down vehicle. That was worse than the crash had been and I started telling them so in some unkind and loud words. I released my belt and climbed out, as did the driver, and we soon discovered that we were not much worse off for wear and tear.

So, that was an interesting part of our pre-deployment training that we didn't want to have repeated during our event that we were now planning.

Everything was going well enough and we were a couple of days away from seeing the Battalion roll into Camp Guernsey and begin to train. Then tragedy struck. One of the support units had convoyed to our location from Ft Carson, including one of the 5-ton trucks that was carrying explosives. As the convoy neared Camp Guernsey, where the weather had taken a turn for the worse and was cold and snowy, the truck slid off the road and rolled upside-down.

We received a call on the radio and rushed down the road to the crash site, which was less than five-miles away. The NCO in charge of the explosives carrying vehicle had been killed in the crash and the junior service-member was at a loss about what to do, other than to call us on the radio.

After confirming the condition of the deceased person, we contacted the State Highway Patrol and our HQ element back at Ft Carson. Due to the explosives on the wrecked vehicle, the Highway Patrol blocked the road and kept all traffic away until the Explosive Ordinance Disposal crew from Ft Carson arrived and ensured the explosives were safe. Then the body, which had been trapped inside the cab, was able to be removed and brought back to Ft Carson.

With the crash and death of the NCO, along with the inclimate weather that had moved in, the training event at Camp Guernsey was cancelled and we soon returned to Ft Carson.

Once at Ft Carson, my Detachment was able to use our various sniper-related ranges under our control to accomplish the training events for the BN and still provide the required pre-deployment training.

As the main body of the Battalion deployed to Bosnia, my Detachment remained at Ft Carson and we conducted a SOTI Course (Level II Sniper Training) and two Sniper Sustainment courses for the Group HQ, among other related activities.

Soon after, I retired from the Army and started a second career focused on Critical Infrastructure Protection, Homeland Security, and various Department of Defense requirements as a contractor. My days of wearing the green wrinkly skin, as my dad used to call his military uniform, were over.

Retired Army Life and then America is Attacked, 9.11.2001

After I retired from the Army, I didn't really have a good plan as far as my next career field or even the choices that were available. I hadn't planned on retiring when I did and had just recently reenlisted for another six-years. But, things have a way of happening in a way that isn't what you had planned. On paper, I was a Team Sergeant of my SFODA. In everyday activities, I was the Team Commander. I didn't have a Warrant Officer nor a Captain assigned to the Detachment. Both ranks, as well as the rank I held, Master Sergeant (E8) were in short supply across Special Forces. I liked my job and since there was a shortage of Detachments that had senior leadership, I expected to stay on a Detachment Alpha, an A Team. My Group Command Sergeant Major had a different idea.

The Group's CSM was at one time, my Company's Sergeant Major, years before and in a different SF Group. We got along and he liked my way of thinking. He was the same Company SGM who placed me on a Detachment that had a weak Team Sergeant and a weaker Team Leader, telling me that they needed some adult supervision. Now, as the CSM of this outfit, he wanted to take me off of my Detachment and place me into a staff position at a Company or Battalion Headquarters.

Granted, I had been assigned to an A Team longer than any other Master Sergeant in Group. But I figured since all the A Teams didn't have E8s assigned to them, I should remain on an A Team as long as I wanted. But, that's not how the Army works, even in Special Forces. He told me to get used to it because I was going to get the staff job whether I liked it or not. I told him I'd retire first. He said I wouldn't. Two-weeks later, I was on terminal leave, headed to retirement. I had three months of vacation saved up and was over the 20-year mark needed to retire, so I dropped my retirement paperwork and went on leave after I quickly turned in my gear and out-processed. I was gone before most people even knew that I was retiring.

About two-weeks before I actually hit the end of my terminal leave and retired, I called the Department of the Army retention office and asked if I could stop my retirement papers and come back to work. I was starting to understand that I had not prepared myself or my potential future position for retirement. I was told that I could stay if I wanted, and all I had to do was get my Group CSM to sign the paperwork stopping my retirement process.

Pride is a great thing, most of the time. I would not ask my CSM to let me stay, so I retired and started to figure out what came next. I bounced a bit, to say the least, and this is what came next.

I worked a short stint as an electrician and was on the road all the time with my working area covering several states, so I quit.

I worked a few months as a car salesman and discovered that I was incompatible with what I viewed as crooks and cheats that made up the majority of car salesmen I met, so I quit.

I had a college degree and I had a Top Secret clearance, but I didn't have a specialty area that I thought I could use as an asset for employment. Which, had I retired in a normal fashion instead of in a huff, I would have known that my military skills were in great demand in many areas. But from my perspective at the time, I felt that there wasn't much call for old Infantry types of soldiers, at least not that I knew of.

Again, that was my fault for not being prepared for retirement. However, personal computers and jobs requiring computer-based skills were really opening up, and computer repair and computer networking jobs were listed on every jobsite. So, I went to school again and added about a year's worth of computer training to my resume.

I became certified in computer repair and in computer network administration. This level of computer training, along with my military security clearance allowed me to get hired as a contractor, working with super-secret "black" networks for the US Air Force. I could do it and did it quite well, but I didn't like it. Still, it paid the bills and I plodded along. And then the Twin

Towers, as well as the Pentagon were attacked by al-Qaeda terrorists who had hijacked airliners and turned them into missiles on September 11, 2001. 9/11 changed everything.

I was a computer network guy and usually I got called to help when someone couldn't get their computer to work or if they locked themselves out of their account. I kept the networks I had assigned to me up and running. After 9/11, I started to get some extra attention from a few of the senior officers who worked in various planning cells during the run-up to our invasion of Iraq. My former job became more interesting to the customer than my current job. I was asked to sit in on some briefings, in case of computer issues, and then I'd be asked my opinions on certain subjects later in an informal debriefing. I was still a computer network employee, but I was providing some requested services that would be best described as "out of scope" for our contract. And I didn't mind at all.

About this time, my own company started to get tasked for Homeland Security issues within the Critical Infrastructure Protection and Homeland Defense arenas. The Department of Homeland Security (DHS) has just been established and was getting started, building itself from the ground up. I was soon given the opportunity to switch from my position of working on networks, to putting my old skills to use conducting vulnerability and security assessments for elements of our nation's infrastructure that were considered possible targets for the Muslim terrorist organizations.

As I transitioned from a Network Administrator to my new position as a Senior Security Analyst, besides a big pay raise, I got to start doing things that were much more interesting. I worked in several major areas of Homeland Security requirements.

One of these contracts was a multi-organizational contract that encompassed entire global transportation routes for various products that traveled between destinations while packed inside Sea-Land Containers. These routes included legs traveled via cargo container ships across the oceans, railroad lines, and big-rig trucking routes, in their entirety from the point of manufacture to the end retail location. Items that were specifically targeted during my operations included such diverse products as food stuffs to automobile parts. My travels took me to places throughout Germany, Japan, and many points from our own ports along the coasts, to distribution centers in the heartland of the USA.

Another area I worked in was fresh water systems, our drinking water sources. This included doing security assessments and working out security solutions for identified vulnerabilities for dams and water pumps, tunnels, and canals that covered the entire Colorado River Basin, from the Rocky Mountains to the California coast, and to the desert areas of Arizona and the Gulf Coast region.

Another huge water system I worked on was the entire potable water system of New York City. This encompassed all aspects of the waterways from the reservoirs and dams, to the water pumping stations and underground controls and conduits inside the boroughs across the city, and everything in-between.

Besides working these various Homeland Security or DHS funded contracts, I was also an instructor for the company's training program on risk management, specifically focused on physical risk of critical assets from infrastructure, to facilities, to personnel. I instructed both military and government employees, as well as some corporate employees. Our basis of instruction was built around what was known as the Continuous Risk Management processes as refined by our company.

This was a structured methodology of defining a specific or general threat as it was weighted against an asset, and then determining the level of risk associated with the asset, based on a graduated plan of mitigation. The mitigation analysis would allow the asset owner, be it corporate or governmental, to reduce or even accept the threat, or "unwanted event" based on various determinations provided in the analysis.

While I was in the military, specifically as a Special Forces Engineer or Intel NCO, I did the target planning for mission folders or operations. As a contractor, I could use the same knowledge and experience to create vulnerability assessments and security plans. I worked in this capacity for several years, and was considered a key contributor for many operations.

SCUBA School with my Kids

After I retired from the Army, it wasn't all work and no play. Whenever possible, I tried to make up for so much time I'd lost with my family while I was in the military. Linda and I would take little mini-vacations around the state, or take small trips for two or three days and sight-see. I also was able to spend some great quality time with my kids, who were quickly growing up and creating lives for themselves. Still, I wanted to stay connected as much as possible. The water helped make that happen.

I would bet that it's a small percentage of fathers who get civilian-rated SCUBA qualified with their son and daughter. I have been blessed by being able to do so. While in the Army and as a Special Forces operator, I always thought that I would have enjoyed going to the Special Operations Combat Diver Course, or SCUBA School, but spending most of my time on a Military Free Fall Detachment had me focused on one specialty skill already, and I could never justify going to the course and taking a slot from someone who would likely be assigned to a dive team.

That stated, once I retired, the thought of diving underwater was still a very interesting concept to me and as I quickly found out, to my son Josh as well. Since this was a goal we discovered we both shared, we immediately set out to make it happen. We began this journey in July 1999 and we both became Professional Association of Diving Instructors, or PADI certified open-water divers. Josh was almost 15 and I was in the process of wondering just what I would do now that I'd grown up and retired from Army. Deciding to become Open Water Divers turned out to be a great thing we could do together.

We signed up for the PADI Open Water Diver training course with a local dive center and got started. After we got our skills down by training in the pool, we went to a place called Blue Hole, NM and completed our certification dives. Blue Hole, located at Santa Rosa, NM is a cave-fed water hole that's 80 feet across and over 100 feet deep in some places. It also connects to an underground water system feeding seven water holes in the area. Shaped like a bell, the hole gets wider at the bottom and due to the altitude being just over 4000 above sea level, all dives must be planned using high-altitude dive tables. It's a great place to get certified, although it comes with its own set of challenges.

My son Josh and I were dive buddies for our dives at the Blue Hole. We were able to experience breathing from compressed air while underwater for the first time, together, both in a pool and in a natural setting. It was very inspirational for me to watch my young son accomplish a potentially dangerous and scary goal, especially while I was learning the same new skills myself. A very proud father was able to share these experiences and perhaps more importantly, some very wonderful time with my son. We never went diving again, other than to practice our skills a few times in the dive school's pool, but I never forgot what a great time it was to become Open Water Certified with my son.

I was lucky enough to repeat the experience of great pride and personal time spent with my son, only this time, with my daughter. Three months after becoming a beginner, I decided to get some of the advanced Diver Training that was available and signed up to become Advanced Open Water Diver certified. My daughter Lori decided that she would also enjoy becoming dive certified and while I was attending the advanced training, she attended the beginner training at the same time. This allowed us to go to Blue Hole together, along with the other members of the two courses. While I was getting my advanced course certification dives completed, Lori completed her cert dives as well. We were able to dive together during the cert dives, but we were unable to be Dive Buddies since we were both still being graded in different courses of instruction. As soon as we were completed with the required graded exercises, we were able to plan and dive our own father-daughter dive.

As a newly certified advanced diver, I was qualified to plan our next dive, which was on October 17, 1999. To make things extra special, Mother Nature decided to snow on us for our first dive together as Dive Buddies. The surface temperature was only 25 degrees and the wind chill was a frosty minus 2. I was a little worried that our tanks or regulators would freeze, or that we would freeze before we were able to get geared up and in the water. Like I mentioned, it was very cold.

It was cold outside, however, the water in Blue Hole was a constant 62 degrees Fahrenheit all year. It was steaming this afternoon as we entered the water and we were instantly warmed up again. We entered the water by doing a backwards fall off the end of the pier. Once in the water and ready to submerge, we descended to 35 feet below the surface and swam around the edges of the cliff and across the depths of the open space for just over a half hour.

Getting back out of the water was painfully cold. It was not only cold, it was beginning to get dark. I froze and put the gear inside the vehicle as Lori wrapped in a towel and jumped in the truck to stay warm. I didn't mind. I was still the dad and she was still my little girl even though she was a legal adult. Just barely, but still 21. I'm sure she enjoyed the legal beer after the dive during supper. I know I enjoyed our time together and enjoyed sharing such a unique experience.

Lori went beyond our dive in Blue Hole and has dived in the oceans with her husband Stew, but I am satisfied, as well as honored and privileged to have gotten to share some extraordinary times diving underwater with both my son and daughter.

New Mexico Counter-Terrorism, Operation Rattler, NMSP and Border Patrol

During the summer of 2004, I began working on a contract with the New Mexico Office of Homeland Security. This contract, named "Rattler" consisted of a requirement to conduct a series of counter-terrorism exercises. These exercises included local, state, and federal law enforcement agencies within New Mexico and were developed in order to bring the state's training scenarios into alignment with, and be sanctioned by, the Department of Homeland Security in Washington, DC. This contract was awarded initially as a three-year contract with the company I was employed by as a Senior Security Analyst.

The initial trip to NM started with a visit to the Governor's office in Santa Fe. There we attended a conference with all the primary members of the NM HLS office, to include key leaders from the NM National Guard, State Police, and NM-based US Border Patrol. Our first Counter-Terrorism exercise was to be a multi-organizational task force working out a plan and then executing it against a terrorist group that had established a training camp on the NM side of the border with Mexico. This would occur in the area known as the Boot Heel in the southwest corner of New Mexico.

I attended the conference with one other member of my company and we were the lead contractors for the entire operation that was to take place over the next few days. As we got ready to depart Santa Fe and travel to the border area, the New Mexico State Police (NMSP) contact, a Captain who was a former NYPD officer, told us to follow him to the training site.

Following the NMSP Captain was a much more interesting story than it first sounds. He was driving a marked police cruiser and we were driving a new black Chevy Suburban rental vehicle. As we hit the highway headed south, he turned the emergency lights on his cruiser and away we went. I quickly found out that a Suburban rental has an engine governor that prevents it from being driven over 100 miles per hour. For the next couple of hours, we maintained the 100-mph speed as we drove south to Las Cruces, slowing only slightly as we whizzed through Albuquerque.

The trickiest part of following his cruiser at such fast speeds was when someone in the right lane would pull into the left lane, in front of my SUV, as it passed a slower vehicle. While most would see the flashing lights of the cruiser and yield, the same was not true for our normal looking SUV with only the 4-way flashers blinking. I quickly figured out that I needed to slow down and be very careful as I overtook slower vehicles in the right lane, even if I was quite close to the cruiser. Some people just don't pay a lot of attention to what's happening behind their vehicle, and I was traveling an average of 25 mph faster than the posted speed limit.

Much sooner than what a normal drive would have produced, we reached the National Guard training area near Cloverdale, NM in the far southwest corner of the NM Boot Heel. The Boot Heel got its name from how the corner extended south, forming an area that resembled the heel of a boot next to the rest of the border. This entire region was also a notorious area for cross-border smuggling routes as Mexican Cartels transported their various drugs and other cargos across this deserted and desolate desert region providing easy access into the USA.

After the pre-mission briefings and final plans had been adjusted, we were ready to start the exercise for the evening. After the sun had set and it was completely dark, I was going to be flying in a NMNG Blackhawk helicopter with a team of Task Force members into an area near a suspected smuggler's location very near the US border, less than a mile north from Mexico.

Flying in the Blackhawk brought back a lot of my military memories, as we were moving low and fast, in total darkness across the desert floor and following the mountain ridges in what is called nap-of-the-earth flying. We landed near the base of a long ridge and started movement via foot-patrol towards a small cabin and fenced area where the Team was to establish surveillance, and eventually conduct a raid. This was in an area named Antelope Wells and was also right in the middle of a real-world smuggling corridor.

Late that night about the time we had finished with the raid against the exercise terrorists, we received radio traffic from our NG helicopter crew that they had been called out of play to support an interdiction of an actual cross-border operation. Apparently, as we conducted our exercise, the other friendly elements assisting had discovered an actual event by coincidence, made possible by monitoring the specific area for the exercise. Bad luck for the Cartel, good luck for the US lawmen.

Soon, local law enforcement had apprehended a dozen or so personnel carrying backpacks of drugs into the United States. I was told that over 1000 pounds of marijuana had been confiscated during this real-world interruption to our exercise.

Eventually the Blackhawk returned and flew us back to the base camp for the debriefing and some much-needed time to rest a little before we got ready to continue the exercise.

As the Holiday Season approached during the latter months of 2007, I got an offer from one of my former Special Forces Teammates for a job in Iraq. The Green Zone, or as we called it, the International Zone (IZ) was going to need a Regional Manager (RM) for a large Intelligence Services contract supporting the Multi-National Forces – Iraq (MNF-I) Command and Staff elements in Baghdad, Iraq.

This was a country-wide contract and included several Regional Managers. My area would be the IZ and a few outlying Forward Operational Bases (FOB) located in the vicinity of the Baghdad area. Additionally, besides operating with the US and Coalition Military Forces, my personnel would also be supporting various US Embassy missions. My experience working with DHS and a few other 3-letter agencies in the Washington DC area was viewed as a benefit for dealing with the Department of State employees working in the US Embassy.

As 2007 progressed into the winter months, I watched several decisions change my business division's direction for future opportunities. I believed that my current position with the company where I'd been working for the last seven years was starting to shift away from Critical Infrastructure security-related work and more towards Washington DC based technical support positions. After discussing the opportunity I'd been offered for the position in Iraq with my wife, we decided that the offer was a good one and one I should take. I then started the wheels into motion.

By the end of January 2008, I had scheduled my last day of work to be during the first week of March. On Friday, March the 7th, I arrived at the CONUS Replacement Center (CRC) at Ft Benning, GA. to prepare to deploy to Iraq. While there, I and other contractor personnel, received our basic military gear such as body armor, medical kits, helmets, and protective mask (for chemical or other contaminants). We also got a couple of days of training in emergency medical care and on how to react to personnel or vehicle borne explosive attacks (suicide bombers), roadside bombs (IED) and various other threats we could encounter during military support operations.

A week later, we flew into Kuwait's Ali Al Salem Air Base, the primary airlift facility for suppling hostile areas in the region with additional combat power. After three days, I, along with seven employees in my company and a C-130 full of military personnel and other contractors headed for the Baghdad International Airport (BIAP). Once we landed and got picked up by some other members of our company, we were taken to the local company HQ, known as The Ranch.

The Ranch was a very rough-cut version of a typical corporate HQ. There were two or three Regional Managers living there, a dozen or so workers who departed for the military customer's locations each morning, and a few administrative workers who managed the vehicle assignments and maintenance, the supply guy, whom I quickly befriended, and a couple of general go-fers.

The Country Manager and the assistant Country Manager worked a few miles away, still on the same Coalition Base that surrounded the BIAP area. The day I arrived at the Ranch, the Site Manager was hosting a VIP Cook-out for the company's VP and the business division's Director who happened to be in-country. Because of their visit, all the Regional Managers had been called into the Ranch's location for the meeting and cookout. This gave me the perfect opportunity to meet everyone, and to find out that I knew several of them already from our time spent in the Special Forces community. All the RMs were former Green Berets, with the exception of one, who was a woman with a Military Intelligence background and a Doctorate Degree in some type of Middle-Eastern studies. She stayed close to the flagpole and the Ranch.

The assistant Country Manager was not only the person who had called me with the offer, he was also a former Teammate of mine from 3rd SFG(A), SFODA-384. We had also served together in a different unit during another period, the 10th SFG(A). Once we linked up, I got a good tour of the facilities and a detailed in-brief on operations, as well as a good visit to get caught up over the next few days. Soon enough, it was time to leave the Ranch and catch a military patrol with a transport vehicle called the "Rhino" headed towards the IZ. The Rhino was an armored personnel carrier that could carry about a dozen passengers, had a turret gunner and driver, and was made to survive most IED attacks. It was what the military and contractor alike used as a taxi to make movements from Camp Victory and the BIAP to the IZ.

During my Intel briefings, I was told about Muqtada al-Sadr, a Muslim cleric who was in charge of the Mahdi Army, one of the coalition's and the Iraqi government's primary enemies. I knew that his militia and supporters were located in a part of Baghdad just outside the IZ on the other side of the Tigris River in an area called Sadr City. This group frequently attacked checkpoints and other targeted locations, and they used a lot of rockets and mortars to reach areas where security prevented them from using troops. IEDs were also used a lot by this group. What I didn't know, nor did anyone else, was that al-Sadr was about to launch a campaign against the IZ using the Christian Holy Day of Easter Sunday as a starting point.

The day I left the Ranch and arrived in the IZ was April the 23rd, 2008. Easter morning.

Welcome to the IZ, Baghdad, Iraq

The Green Zone. A city within a city, isolated behind large concrete barriers called T-Walls and guarded gates that provided just slightly more than a false sense of security from the dangers of being in a war zone with no front lines.

I had roughly 50 employees working for the G2, or Intelligence element of MNF-I, led by a 2-Star General who worked for the 4-Star Commander, General Petraeus. As opposed to all my contractors, I didn't have a military supervisor. My position was strictly a company position and I made sure that my people were fulfilling the requirements of our contract with the customer, the G2 of MNF-I. I was considered the equivalent grade or "rank" as a government employee rating of GS-15, which was a parallel rating to the Army's rank of Colonel.

The first thing I did as I was greeted at the Rhino's dismount point, was to request a tour of all work locations for my employees. I needed to get a quick idea of the layout of the IZ and how to get around. There were several FOBs located in the IZ and I had personnel working in all sectors and quartered in several different locations. First, there was the US Embassy compound. The Ambassador and the State Department employees worked in one wing of the former Iraqi palace. MNF-I HQ and an assortment of support elements were located throughout the remaining areas of the palace, which encompassed a basement level, three main floors, and a fourth, much smaller floor with access to the roof. It was on this top floor where I eventually located my office until the coalition departed the palace and moved into the new US Embassy compound a couple miles down the road.

Also in the IZ, were FOB Union III, although Union I and II were no longer in existence, and FOB Prosperity. This area was out on the edge of the IZ and functioned almost independently from the rest of the Green Zone. Camp Olympia held several contractor security and non-US or Iraqi military units that were part of the coalition. FOB Blackhawk extended into a different direction and contained the Al-Rasheed Hotel and the primary pedestrian gate into the IZ for local people seeking employment or to visit the Iraqi governmental offices within the IZ.

At least half, if not more of the area that made up the International Zone was Iraqi national occupied. There were multi-story apartment buildings, homes, private compounds, businesses and restaurants, and various other organizations and personnel who were living there when the Green Zone was mapped out and declared a security zone. Normal people without ties to the enemy were permitted to stay in their homes and businesses and simply had to get ID cards or papers which allowed them to leave and return as required. The risk of someone with authorization to be in the IZ, turning out to be an enemy combatant was real and occurred regularly.

The contract with my company was valued at over $500 million and within the IZ itself, I had over $500,000 worth of equipment, such as vehicles, computers, and electronics that primarily supported administrative requirements for the employees. Most mission-related requirements were provided by the customer.

The locations of work sites were as diverse as the contract specifics for skill sets. I had people working in the IZ enemy detention center, the US Embassy, various military worksites, and a few stand-alone contractor-only worksites. Risk management for the safety of the worksites and the company personnel was one of my primary responsibilities, especially if the site physically changed location.

As I was shown the different sites, my first introduction to the Big Voice occurred. The Big Voice was the emergency warning that sounded when a rocket or mortar attack was taking place. It echoed across the entire IZ and was a combination of an alarm and a recorded broadcast stating in English to take cover. As I heard this alarm, we had just entered a compound where I had three or four people working. We parked and entered the office area which was considered cover due to the thick block walls and roof. I quickly found out that this was the first time the alarm had gone off in a couple of months, and I was asked if perhaps someone on the enemy side might have been irritated to see me in their own backyard. An obvious joke, I still assured them that I was not responsible for the attack.

Rocket's Red Glare, Bombs Bursting in Air

Twice during my trip around the IZ we experienced rocket attacks and a call to shelter. Something was going on, but having just gotten on the ground, I hadn't noticed anything out of the usual, since it was all unusual at the moment.

Once I'd been around the area and checked my map, I dropped my escort off to go back to work and then drove my new-to-me pickup truck to my hooch area. My vehicle, for the first couple of months anyway, was a small 4-door pickup truck. I drove it onto the US Embassy compound, which was where my housing area was located. At the time, the US Embassy was located in the old Iraqi Palace where Saddam Hussein lived and worked, at least part of the time.

The housing for military and contractor personnel was created by four villages located around the palace grounds, each filled with small trailers. The trailers were called CHUs (pronounced choo) for Containerized Housing Unit. These were about 22x8 feet, and generally held four personnel. My CHU was split in the middle by a shared latrine with a sink, toilet, and shower. I lived on one side and an Army Colonel lived on the other side.

I had parked at the end of my line of CHUs that was one of about twenty rows, making up what we often called CHU-ville. My area was named Edgewood and I was in the last one in a row, next to an internal barrier wall. I noticed that mine had a row of sandbags stacked along the front, but that most of the others didn't. Since mine was the end CHU in the row, it probably had sandbags stacked along the front in case something blew up near the barrier wall. That, or one of the last occupants had been nervous. I was about to become thankful for those sandbags.

I had been in my little trailer about ten minutes and was unpacking my bags when a huge boom sounded as simultaneously, I felt the CHU jump up and down an inch or two and I was showered with dust from the ceiling and walls. And then the Big Voice went off. My ears were ringing as I wondered what else was going on, whether we were under any type of assault or whether it was a single rocket or mortar. The dust and smoke outside my CHU hung heavily in the air as I looked around the area for anything or anyone that might need some attention. I had quickly thrown my body armor and helmet on and figured I would stay near my sandbagged corner of the compound until I got some additional information or saw something to adjust from.

While I looked outside as the smoke and dust cleared, I saw that something had exploded about ten-meters from my trailer. It had detonated on the doorstep of the trailer next to mine and covered everything with shrapnel as well as gravel from the ground. My sandbags had caught everything that had been headed into my trailer and prevented me from possibly becoming injured, or worse. However, the two CHUs nearest the blast didn't have any sandbag barriers, and they had a lot of small holes in their aluminum skin.

Since it hit about lunchtime on Easter Sunday, many people were either eating in the DFAC, (Dining Facility) or were working. Every day was a work day in the IZ. However, there had been one guy, a US government civilian, in a CHU in the row next to mine and without any sandbags protecting his exterior walls, he got hit by shrapnel and died from injuries during the attack. He didn't yell, at least not that I heard, and I didn't know he'd been killed until after he'd been discovered later by his roommate.

I found out later that the IZ had been hit with multiple 107mm rockets that had been fired into the secured area by the Mahdi militia from Sadr City. Several personnel within the IZ were injured in the attack, but only the one American was killed that day. And as luck would have it, the one that got him almost got me, and on the very first day I was in the IZ. Welcome to Iraq!

I quickly discovered that there had been almost no method of accountability established between my employees, outside of their own military counterparts. Whether it was communication, equipment, or even standards of conduct, I found out that much needed to be tuned up. As it also turned out, the attacks I experienced on Day 1, were the beginning attacks of a six-week campaign against not just the IZ, but against MNF-I targets across Iraq. As stressful as the daily attacks were, sometimes several within a few hours, it added emphasis to my implementation of new procedures for status updates and regular accountability requirements I decided needed to be established. I had hit the ground running and was not going to slow down for a long time.

We continued to get rocket strikes in the IZ daily. A huge sand storm blew into Baghdad and the Mahdi Army used it as cover to attack the perimeter and several checkpoints. A week after I arrived, I had one of my outstations hit by rocket fire. One employee was slightly injured, some equipment damaged and a vehicle received shrapnel damage. I was able to respond and assist the employee, as well as recover the classified equipment that had been damaged. We got the site operational again within a couple of weeks. During this time, all personnel in the IZ travelled with body armor and helmets whenever outside hardened structures that could provide cover. Morale was low among the contractors and I was visiting each worksite daily to ensure that security was in place and to alleviate fear or concerns where possible. I had quite a few personnel decide to terminate their contracts during the 6-week period of almost continuous attacks, which were then replaced by new personnel. This constant rotation of personnel kept me very busy with transportation and in or out-processing requirements. It really didn't slow down until towards the end of May.

As if all of this was not enough, as I quickly discovered upon starting to work in the IZ as the RM, I was basically creating a new role to be filled as far as the military customer was concerned. Apparently, there had not been a designated RM in the IZ for several months prior to my arrival. After asking around, I found out that the former RM had done very little besides stay out of the way and leave everyone alone. His primary function had been to schedule transportation into and out of the IZ for personnel as required, and to pick up mail for several of the employees who worked at outlying stations around the IZ. The majority of time, he stayed in the Embassy's café drinking coffee and using the unclassified wireless network for any computer work he completed.

One of the first things I did was to request office space inside the Embassy and to get a government work station set up for both unclassified and for classified computer networks. I had to explain my contractual function to the government's contracting officer and then physically search the offices that my company supported for open desk space and adequate equipment. Within a few days I was set up and operating, and by now the customer also was aware of how I could benefit the entire process from employee issues, to business development. One of the first issues the military asked me to work on was a request to place some of our company people into an area outside the wire, but not quite in the Red Zone. It was my 1st introduction to the Pink Zone. The Pink Zone was a semi-secured area that really didn't meet the definition of my company's requirements for workplace security. It was an area physically outside the IZ or Green Zone, not controlled by MNF-I coalition troops, but rather, secured by Iraqi troops or police and was not considered to be in the Red Zone.

Pink Zone Security Assessment, Adnon Palace, Baghdad

The IZ had a large Iraqi government building located inside the secured perimeter. This building was further secured by Iraqi guards and checkpoints and the building itself was almost entirely occupied by members of the Iraqi government for the city of Baghdad and the entire country. I would accompany one of my contractors into this building when he needed to be there if he didn't enter along with other US military personnel.

The building was considered secure, but because of the total Iraqi control, the contract security posture for the company employees was considered very minimally fulfilled by this location. I went here every couple of weeks, but none of my personnel worked there all the time, even though it was considered a "green" area within the IZ.

The Adnon Palace was different. Near the infamous Crossed Swords parade field inside the IZ, the Adnon Palace was located outside the US Forces secured area of the IZ, while still not being considered in the Red Zone. This facility was secured by an Iraqi manned checkpoint, but essentially, anyone from the Red Zone had access to the Adnon Palace without being checked by US and Coalition personnel.

During this time, the Adnon Palace was used for a lot of local resources such as licenses, permits, and other requirements involving local government and military officials outside the roles requiring MNF-I interface. There was, however, one small office manned by US military personnel inside the palace for communications and liaison requirements that were not conducted from within the IZ proper. This office was described as a Command Information Center (CIC) and was an extension of the Baghdad Operations Command (BOC). It was described officially as being established "to better facilitate combined operations planning, coordination and intelligence analysis at the national level of operations command." The customer had asked our company to provide some contractor support and I had to determine the security level and whether or not it met the requirements of the contract for our personnel security.

During my assessment of the location, I discovered that I could walk freely to any of the multiple levels of floors, to include the basement. I could also move outside the palace and observed several paths that crossed into the palace grounds from the Red Zone without any checkpoints or static security positions. This opened the palace grounds and even the interior of the palace itself to unauthorized personnel who could be in possession of weapons or explosives. I had to refuse to allow any contractor support to be assigned to the CIC as the contract was currently written.

During my time in the IZ, I had to make several security assessments for new locations where the MNF-I customer wished to place some of our employees. The Adnon Palace was the one location that I was unable to authorize as a workspace.

FOB Abu Ghraib, Supply Depot Security Assessment

Abu Ghraib was infamous for having a large military prison located on the Forward Operating Base (FOB) where several US Military Police got in trouble for abusing Iraqi prisoners. FOB Abu Ghraib also contained a large supply depot and had a very small number of military personnel and a few local-hire employees who worked to safeguard and maintain accountability of the equipment and supplies stored on the depot grounds.

Located about 35 kilometers outside Baghdad, the FOB was surrounded by the Red Zone, although it was only about 15 kilometers from Victory Base Camp (VBC) and the BIAP area. I traveled to this FOB to complete a security assessment as they had requested several of our company's personnel to do a temporary visit to interview more local-hire personnel to fill an increase of positions inside the supply depot. One of the largest efforts our employees filled was security vetting for local people wishing to get a job working on a military or US Embassy location within the IZ.

I had to use a private security team to travel from the IZ to Abu Ghraib. The PSD, or Personnel Security Detachment was provided by a local company called Falcon Security. My patrol leader was a former soldier from England, the medic was also a former soldier who served in the South African Defense Force, and the security team members, the shooters, were all local Iraqi personnel. Most likely former Iraqi soldiers or perhaps converted terrorists. There was no way to really know who they were for a private company. I rode in an older white armored Suburban with the patrol leader, who also drove, and the medic. The crew rode in the back of a pickup truck with a machine gun mounted above the cab. I was told if we needed it, that there was an AK47 and some full magazines in the rack by my seat in the vehicle. I didn't get the same sense of security that riding in a US military convoy, inside a US manned Rhino gave me, but it was exciting, nonetheless.

Our route to Abu Ghraib took us through some seriously rough looking neighborhoods. At any moment as we travelled along the roads and streets, I expected to come under fire or to hit an IED. Many of the buildings, former homes, were piles of rubble having been destroyed during fighting in the area. Some were being lived in, even if portions of the houses had been destroyed. Laundry was hanging from the lines stretched across the open areas and there were still some kids playing along the streets. Life always goes on, at least until it ends.

Soon, we were approaching the entrance to the supply depot and I left the PSD and entered through the security portal. I met the Commander of the complex as I passed through the entrance and was greeted as if I were a long-lost friend.

The site Commander was a young USMC Captain and he apparently didn't have a lot of people he could speak with under his command. There were few military members, but mostly civilians from what I had seen. We went to the building that held the dayroom and several hotel-like rooms as he showed me the suite I'd be staying in for my visit. It was much nicer than I'd seen so far, with its own bathroom and a large living area. The windows were barred and the entire building was ringed by T-Wall sections that overlapped each other by a foot or so. Once I dropped my rucksack in the room, we went to his HQ office and started getting busy.

At his office he offered up an assortment of weapons for my use, should we need to defend ourselves. I asked him the last time that had happened and he said it had been a while and that he was just kidding. We discussed the reasons he wanted to increase his work force and how long he expected the screeners would be needed as he held interviews. Then we toured the area which was a huge storage yard surrounded by high concrete walls and topped with razor wire.

Having the infamous name of Abu Ghraib, I had to ask about the prison. It was located a couple thousand meters from the supply depot and there were no former detention facilities at this location.

There were a dozen or so large warehouses and rows upon rows of brand-new civilian vehicles. There were literally hundreds of new F-150 and F-250 4WD trucks, civilian ambulances, several buses, and dozens of SUVs. The warehouses contained all the calibers of ammunition needed for military weapons of both US and other types, new and used rifles, pistols, and machine guns. Even hand grenades, mines, and explosives. There were a few guard towers, but I didn't see any guards manning the towers anywhere around the perimeter. The only place I'd seen guards was the front gate where I entered the facility. There had to have been millions and millions of dollars of equipment stored there and I wondered what the final disposition of it all would be, whether it would be used for the purposes it was intended, and if any of it would even be returned to the USA.

I have to admit, when I went to the dining facility to get supper, it was a bit odd. First, it wasn't set up like the typical DFAC I'd eaten at on any other FOB or Camp in the IZ or at VBC and BIAP. It was set up much more like you would expect a strictly Iraqi café to look. Likewise, of the 50 or so people inside the DFAC eating their meals, there were only a few American military members. It was filled almost totally with local-hire employees, some of whom were workers and some were security, with uniforms and weapons. It was much more local-flavored than I had expected, but I was not treated as an outsider or with any apparent hostility. And, the food was good.

I finished my assessment towards the middle of the next day and made my way back to the IZ via the Falcon Security PSD again. I often wondered how the Marine Captain did as he finished up his tour. He was like a king of his own world from what I could tell, and seemed to like the job. I'm sure that it was a fairly easy job, until or unless something really bad happened, and then he would be on his own, at least for a while.

Once I was back in the IZ I submitted my report, recommending that we allow the screeners to go. The company and the customer scheduled a 4-day mission and it went off without any issues.

The Holidays in Iraq, 2008

The Christmas Season in Iraq included a US Presidential election and George W Bush was replaced by Barak Hussein Obama. Things were about to change for the ongoing efforts in Iraq and more than the US was watching.

About Thanksgiving, the IZ started getting a lot of Intel warnings about bombings and indirect fire attacks. Vehicle-Borne IED bombings had increased across Baghdad and the security posture within the IZ had been increased. However, the US security on the gates coming into the IZ had been replaced by Iraqi security. Having the Iraqis in charge of the searches and control of all vehicles entering the IZ was not a welcome change for security. About this time, we were also being told that the Rules of Engagement were being changed, making it harder to shoot at suspected threats. The focus was now going to be on what was called de-escalation, where instead of shooting the enemy, there would be required steps to stop violence before it happened with non-lethal means. In other words, don't shoot unless you were being shot at, and you could positively identify the shooter as a bad person.

On top of that, we also got new rules for contractors who lost their diplomatic immunity when we also lost an approved Status of Forces Agreement between the US and the Iraqi government. Legally, any US civilian or contractor could be arrested by the Iraqi police or military and the US military would not legally be able to do anything to prevent it from happening. This made any military patrol with contractors in attendance, much more interesting.

Christmas Eve brought at least two large VBIED explosions on the perimeter of the IZ with one of them being close enough to my position to rattle everything for several seconds and make my ears pop with the concussion. Christmas Day, one of the dining facilities was hit by a rocket and the roof collapsed. Then on New Year's Eve, there were so many weapons being fired into the IZ from the surrounding Baghdad neighborhoods, it seemed like it was raining bullets. Meanwhile, the IDF (Indirect Fire) of rockets and mortars continued throughout the entire Holiday Season.

One big change over the holidays was created by the move of the US Embassy from the old Iraqi Palace into the newly constructed facility that became the largest US Embassy building in the world. As a part of the move, once the State Department and the Military had cleared their respective sections of the old Embassy, I, along with three or four members of my personnel who ran the Special Security Office conducted a final sweep of the entire facility. The SSO, my people, were responsible for managing the classified networks and work areas, which included ensuring all classified material had been removed prior to turning the facilities back over to the Iraqis.

We went room-to-room, checking every desk, file cabinet, crook and cranny for documents or anything sensitive that may have been left behind. This required us to physically remove all drawers, which usually meant breaking or destroying the furniture that had been left behind.

We found a lot of sensitive documents in our sweep, and even found some classified information that had been physically, lost between the cracks. I made a sweep of the former medical station and found that the drug cabinet had been left fully stocked. The storage area remained filled with medical gear and drugs, to include narcotics. I boxed it all up, to include vials of morphine and other controlled substances, and transported it all to the military hospital. It had been under the control of the State Department staff, so I figured the military could always use it as a bargaining chip.

The entire complex was surrounded by T-Walls and handed back over to the Iraqi government on January 1, 2009. I don't think there were more than three or four desks in the whole place that didn't have at least one broken drawer when they got it back, but I was also sure that there was not one document remaining that should not have been there.

Raising Old Glory over the US Embassy and former Saddam Hussein Palace

Before the Military and State Department left the palace, I wanted to make some special gifts for my family back in the USA. The US flag flew every day from a flagpole in front of the former palace during the occupation, but not from the top of the very large complex itself.

Above my office, which was located in the center and at the very front of the palace on the small fourth floor, directly adjacent to the huge copper dome above, there was a sandbag bunker located on the roof. The bunker had been there a long time and had been used for probably several reasons, but it would provide cover for anyone on the roof working with the many antennas, communication equipment, security cameras, IDF equipment, or even laying out to get a tan.

I took several US flags I'd purchased at the PX near the Embassy and flew them from the roof of the palace while standing on the bunker. For the few minutes that these few flags were flying, they were the highest-flying American flags in Baghdad. After they had been flown in the Baghdad breeze, I made a certificate for each of the flags stating when and where they had flown, and then I sent both of these items to my family members as mementos of the Iraq War and the time I served in support of Operation Iraqi Freedom.

Going Home, 15 Long Months

My tour was originally a 12-month gig, but the company gave me a nice bonus every 90-days. After 2009 started, the ongoing threat levels continued to elevate and a lot of people decided that it was time to leave the contract and go back to the USA. The more people leaved, the more I was asked to consider staying. About the middle of January, I had decided that I'd stay until June and get the extra 3-months of pay and one more bonus. Plus, I'd found out about that time that Josh and Annie were going to have a baby around the end of May and I wanted to be home for my first grandchild's birth, or soon thereafter.

The winter in Iraq was about as nasty as the IDF and the constant stress of being in a combat zone. Every few days the dust storms would roll in and it would literally rain dirt from the sky. It was a bright orange dust and it burned your eyes and nose as you breathed it in, even if you wore a face covering. Besides the weather, the status of the company's contract started to become annoying. Our status as contractors was in the process of being worked out since politically, we no longer held the same rights as we did prior to the now expired, Status of Forces Agreement.

Also, the Iraqi government was in the process of accepting large portions of the IZ from the MNF-I Forces. Besides the US Embassy moving to the new facilities along the Tigris River, much of the work space that had been used by the military and by employees on our contract had been returned to Iraqi control. I was in the process of relocating more than half of my employees to not only new work spaces, many of which had to be rebuilt, but I had to arrange for new living quarters as well.

With the numbers of personnel becoming greater on a smaller number of IZ FOBs, the threat of injuries or death from incoming rockets or other weapons was also increasing. By June, I was more than ready to leave Iraq and return to the world.

My trip home required a two-day stopover in Kuwait while awaiting a seat to fly back to Ft Benning. Once at Benning I stayed only one night to out-process and then caught a flight back to Colorado.

When I arrived at the Colorado Springs Airport, I had been awake for about 48-hours or more with only a little nap on the plane. Still, as I walked out of the terminal towards the baggage claim, I saw my family waiting on me and Josh and Annie had a little extra bundle as well.

I met my first Grandchild Avalyn at the airport. She was all of two-weeks old and quickly, like in two seconds, won my heart. I decided at that moment that I was done making a living where I was going to have people shoot at me. I was ready to settle down a bit more and start enjoying being a grandpa. Desert Storm, Bosnia-Herzegovina, and now Iraq were all behind me and I had no plans to head into a combat zone again. I was glad to be home.

The Retired Warrior, Post-2011

My life was not finished by any measure, but I had reached a point where I suddenly turned a corner and found that the road ahead had flattened out and became much smoother. When I returned from Iraq, I returned to a different set of variables and expectations. In many ways, I had achieved the pinnacle of success that so many have strived to reach and yet, somehow always fell short. This was a new, if not a totally unexpected life-adjustment for me.

During my adult years, starting in my mid to later twenties, I began setting goals based on my personal and professional desires. I had no real long-range goals exceeding the time it would take to watch my children grow up and become self-sustaining adults, but I seemed to have new short-range goals that added to the list all the time. My personal and professional goals all revolved around my immediate family and my military career. As for retirement, I never considered a life after the military and my dreams never reached beyond being a soldier-father-husband. I always tried to put all of my effort into the task at hand and believed the old cliché that if it was worth doing, it was worth doing right. I joked during the military that I never figured I'd make it past 50 anyway, so why worry about retirement.

While I had many goals over the years, I didn't reach them alone, and I enjoyed the support of my family as I pushed to make our lives better with each new opportunity. I wanted to be a good example as much as possible and live my life in a demonstration of what I encouraged my children to do along the way. I liked the idea of not only talking the talk, but walking the walk.

I wouldn't rely on luck for success, other than hoping to be lucky in situations that were completely beyond my, or anyone else's control. Being prepared for an event was important. It was also hard and even impossible at times, but a little paranoia on my part always seemed to go a long way in keeping me out of trouble or harm's way.

Seemingly all of a sudden, I found myself at a stopping point in my relatively short period of what some call the productive years. I got my first job in 1972 and worked a full-time job until mid-2009. Some say 37 years is a long time, but it passes really fast when you're having fun. After returning from Iraq, I worked part-time on several short-term projects as a security consultant. I had quickly discovered that the type of full-time work I was willing to accept, or that I would find interesting, would require me to either relocate to a different region of the US, or even a different country. Based on the amount of time I was willing to be on the road or away from my family, of whom all were located within 12 miles of my house, my options for a full-time job were unacceptable.

My priority was now to be near my family and I was willing to neither relocate nor do much traveling. Not within the US, and surely not back into overseas areas considered combat zones. I had grandchildren now, and grown children, and I had left my wife alone enough times already. So, the short-term projects would have to suffice.

I soon found a few opportunities through my former associates. I was asked to participate in a couple of military studies being conducted for the Army that were looking into updating and improving operations conducted in underground environments, such as caves and tunnels. This included diverse regions and usages such as found in Afghanistan, or along the border between the US and Mexico. Another study I worked on was the growing problem with what was called Green on Blue attacks. This was when a foreign allied force such as the Afghani military would have individuals ambush their US counterparts, usually from within secure zones.

Another short-term project I worked on was called Operation Mountain Guardian, a Department of Homeland Security directed full-scale exercise and evaluation on how the city of Denver, Colorado would conduct a multi-jurisdictional and multi-agency response to a city-wide series of coordinated terrorist attacks. I worked on this exercise as a planner initially, and then as the Operations Chief for one of the exercise locations which included multiple targeted facilities. This project was completed after the event took place, was evaluated, and various reviews were conducted, to include identifying issues and developing recommendations for improvement. Then it was over.

Finally, I was contacted by a company to write weekly articles on home and personal security advice and information. These articles averaged 500 to 750 words apiece, and covered everything from how to maintain situational awareness during everyday activities, to home security systems, to how to prepare for a national disaster.

This continued for a couple of years before the company dropped the advisory program.

That was the point where I decided to retire completely and focus on whatever popped up at any given time. I still figured that after making it through some of the tight spots I'd found myself in as a soldier, and then again, as a contractor, everything from this point on was gravy. I was going to try to enjoy my so-called Golden Years as much as possible while I was still young enough to enjoy them. I have since enjoyed and continue to enjoy, not having a supervisor, nor having anything more than a self-imposed deadline for something I decided to do. Most of my projects quickly became focused on grandkid-related activities and jobs around the house, on the property, or some other event that I found interesting.

Life was always good and it remained good.

Five Years, Five Grandkids, 2009 – 2013

The baby years. What a wonderful and previously unknown and even, unexpected joy. As a grandparent, the baby years were simpler then they had been as a parent. Shorter and less tiring, that's for sure.

I clearly remember telling my kids as they grew up and started dating just how things were going to be once they married and had their own kids. I'm not going to be a built-in babysitter I'd say. Your mom and I are going to be doing our own thing once you're grown and don't live here any longer. Don't expect us to be saddled down with grandkids while you're out running around having fun. It'll be our turn then.

Man, was I ever wrong about how I would feel once the grandkids actually started to come into our lives. Before grandkids, I was happy enough and didn't feel like I was missing anything. I had great kids. Both apparently listened to me while they were growing up as I extolled the virtues of having a functional education and being started in a career by the time they reached the ripe old age of 25.

My children were already successful adults at a young age, an age when so many grown-children are still living in their parents' basements and not able to earn a living with their almost worthless, but terribly expensive college degrees.

Likewise, my daughter and son were married to equally well-positioned and intelligent spouses, who made our family that much stronger by their becoming our in-laws. I don't feel like it's bragging if it's true. I had great kids, by blood and by marriage!

My wife, she was also great. Still is. A successful Registered Nurse who also maintained the home fires and raised the kids while I traveled the four-corners of the earth and did all the things a professional soldier does. Between the kids and the wife, I was happy. But when the grandkids came along, that opened a door I didn't know existed.

I fell in love with my grandkids and suddenly understood at least one of the hidden meanings of life. Happiness is experiencing your successful family be fruitful and multiply. Something perhaps, I should have more-fully discovered as a new father, but somehow my job and other activities managed to interfere with my total appreciation along the way.

For me, I felt I was getting an almost impossible opportunity to enjoy once more, so many of the cherished events from my days of watching my children grow up too quickly.

The Christmas of 2009 was special in a few ways.

I had spent Christmas Eve and Christmas Day of 2008 inside the Green Zone of Baghdad, Iraq. I worked both the 24th and the 25th, well into the night. We were being provided unwanted gifts from the enemy in the form of rockets, with a few car bombs thrown in for good measure. Not a normal celebration, and the first that I had ever not spent at home.

Christmas 2009, however, was much better. That Christmas Eve saw six big stockings hung across my and Linda's fireplace mantle, and one small stocking for Avalyn, our first grandbaby. A baby really makes Christmas special, and for each of the next four Christmases, I added another stocking to the group.

2010 brought Kody, 2011 Harper was with us, 2012 Kerrington was just a little thing, but she was there, and 2013 brought along Sam. Eleven joyful stockings hung by the chimney with care. And each stocking is filled to the brim with memories still enjoyed by a thankful grandpa, dad, and husband.

My kids, now parents themselves, may not fully understand how great it has been to watch their children grow and learn the things that to me, become flashbacks to a time that they were growing and learning the same things. More often than not, I had to miss the first iteration, but watching my grandkids and sharing these experiences with them, allowed me to feel even closer to my own children as the new and old memories almost blended together at times.

Love doesn't get divided, it only multiplies. The joy of being a grandparent, at least for me, was a real beginning to a new chapter. I know it's not that way for everyone, and in some cases, it can't be that way due to circumstances beyond our control. But for me, in spite of, or because of anything and everything that influences what has happened or whatever happens in the future, my grandkids filled a place I never even knew existed. And that's a truly extraordinary tale.

Tidbits and Insights
Memories of My Grandma Perry

I first remember my Grandma Perry as the woman who had such a neat house way out in the country. It was a house with a fireplace where we could actually send letters to Santa by letting the heat suck them right up the chimney. It had to have worked, as I never found a letter lying outside in the snow and we always got presents.

The fireplace also had an eternally cool cement hearth that held footprints in one corner made by my mom, who I found hard to believe when I was one, but used to be a little kid. It was like a part of the sidewalk in front of the Chinese theatre in Hollywood. Grandma never got tired of telling me about how my Mom was once little like me.

There were so many unique items on the shelves behind her chair that I could always play with, that normally would rate a scolding for even getting close to, let alone touching. Little brown Indians, animals, tiny plastic babies, a postcard with a real piece of cotton on it, seeds and all, and lots of brightly colored objects. I can't remember what everything was, but it always looked bright and it was always so special, playing with all the things on her shelves, behind the chair, next to the steps.

My Grandma would even let me play with her poker chips whenever I wanted to, and she had a million of them. Of course, I would have to keep mine separate from the grownups' chips whenever they were playing cards around the dining room table.

Food always tasted so good at Grandma's. I think it's because I could eat on my own folding tray table that she kept at the bottom of the stairs, just for when I would come and visit. She would always make a few extra homemade noodles so I could eat a couple of them raw. She knew I liked them like that.

I even enjoyed watching the shows on her TV that I didn't watch at my own house. Shows like Red Skelton, which she always called Red Skeleton. Shows like Porter Wagoner (when Dolly Parton was a young girl), The Grand Old Opry, The Jackie Gleason show, Daniel Boone, Bonanza, Have Gun Will Travel, Rawhide, Bat Masterson, and probably most of all, Walt Disney on Sunday nights. I don't think she ever made me go to bed before I was ready, and when I was ready, she would go to bed with me and scratch my back until I was asleep. It was good to be her Grandson.

That was how it went for many years, visiting whenever the Army allowed my dad to take us for a week or two to spend visiting her. Then, after life wove some of its tangled webs, my mom, my sisters, and I moved right there on her property. I got to see her every day. She let me drive her car and I learned how to operate a standard shift automobile all on my own. She would turn that old 1961 Rambler over to me as soon as we hit Luhrig Road as we were coming back home from town. She let out a few squawks when I tried to start off in high gear, or when I tried to stop without pushing the clutch in, but never made me quit, and I finally learned how to drive and shift gears all on my own.

As I got older, I moved into her house with her and we shared a couple of years as the original "odd couple". She never griped about the hours I kept between work, school and dating my future wife, she was just there to support me whenever I needed it. I'd get home and let her know I was there and she'd go back to sleep, or wake up from having fallen asleep while watching TV and go on to bed, "now that your home and safe" she'd say.

As I grew older, became married and traveled the world with my own Army career, I always welcomed coming back and seeing Grandma. She never had a hard word for me that I knew of. She had a thing or two to say about a few people once in a while, but only in passing and not in a confrontation. Grandma was a person who made you feel good.

The last time I saw Grandma, she thought I was my Dad. She figured it out after a while, but she had begun to live her life over, in her mind, in a place and time much more to her liking than what she had endured during the last few months. I believe she truly enjoyed her life, and to the best of my knowledge, there was not anything she didn't get to do that she had wanted to do.

Before passing, she was aware of her faith and the promise it brought with it. I know she was not afraid and would not want us to grieve for very long. She is with her Lord and enjoying all of heaven's gifts. We all miss her, but we are much better for having known her and having been loved by her.

Memories of My Grandma Easterling

I was the outright and obvious favorite grandchild of my Grandma Easterling. No doubt I reminded her of my dad when he was a kid and I was giving her a chance to relive her youth in a way I never understood until I started to spend time with my own grandchildren.

Whenever I got a chance to visit Grandma, it was like visiting one of my friends. Grandma would play with me as well as, or better than any of my regular friends. We would get sticks and use them as rifles so we could go hunting for bear. We would always be watchful for Indians while hunting, because we both knew the woods were full of Braves hiding behind some tree just waiting on us to come along so they could spring a trap. And Grandma could hike in the woods as good as anyone. We could climb to the tops of hills that allowed us to see into the next county. If I visited her at home, because she lived in town, we would walk everywhere and even take shortcuts behind stores or across vacant lots where most grown-ups would never consider walking.

I loved eating at Grandma's house. She always had crispy fried potatoes for dinner. And usually beans and pork chops, but I remember the potatoes the most. Crispy brown and tasting like French Fries. I remember she would always say that she couldn't eat without bread. Bread and butter were always on the table.

She had an AM radio above the sink and it was on from the time she got up until she went to bed at night. Grandma never had a TV when she lived in this house, the one where my dad grew up. Later, after Grandpa died, she moved to an apartment and she got TV then. But while she lived here, on Fern Street, or as it was called locally while my dad was growing up, Angel Alley, she only listened to the radio.

She also played solitaire. She would literally, wear the ink off of a deck of cards by playing so many games of solitaire. I know she knew at least two ways to play, because she taught them both to me. And she could build card houses also, and shuffle cards like nobody's business.

Grandma was also a puzzle genius. She would put together jigsaw puzzles with over 1000 pieces and it wouldn't take her all that long. And then, she'd put the pieces back in the box and after a few more puzzles, she'd make it again. She always had a puzzle she was working on laying on a large piece of cardboard either on, or underneath the spare bed. I'd help her for a while, but I never found the joy in spending much time on a puzzle. I enjoyed playing, and apparently, so did she.

When I was about 14, I got a minibike. It only worked about a month before it quit starting and I don't know what happened to it after that. But for a while, the minibike worked and Grandma let me ride her around on it. She'd be screaming and hollering the whole ride. I don't know if she liked riding, or if she rode behind me because I asked her to go, but she always went. My other Grandma used to call her Fear Not. I can see why.

During my break in service, I started to get closer to the Christian beliefs I'd been raised with and spent a year studying the Bible and learning about Christianity. During this time, Grandma spoke with me several times about her own faith and some concerns she had about her status as a believer and what might come next.

She soon decided that she wanted me to baptize her and complete something that she'd wanted to do for a long time. In a pond behind my and Linda's home in Albany, I proudly baptized Grandma and fulfilled her wish. She wasn't a church-goer, she wasn't what you'd call religious, but she believed in God and that Jesus was her Savior. She was now comfortable with the concept that her next level of existence would not be made worse by failing to follow what she considered a critical step in salvation.

Grandma died several years later, after I had returned to the Army. I was able to attend her funeral and traveled to Athens from Ft Bragg, NC. I was sad, but I also was very happy for the time we had gotten to spend together. More a friend than a grandma, she helped create some of my most cherished memories as a young child. The visits at Christmas and during my dad's vacations while he was in the Army always included trips to see my Grandmothers or, they would come to us and visit.

Lucky for me, my Grandmothers were also best friends, and usually when you saw one of them, the other one was close by. They shared a very special family bond that never came apart when my mom and dad divorced. I'm very thankful for the time we had together.

Spirit Walk Poem

I wrote a poem after retiring from the Army in the fall of 1999 and on a lark, sent a copy to an organization called The International Library of Poetry. Not long after, I received a letter along with my poem outlined in something called a “publisher's proof” telling me that my poem had been selected to appear in an upcoming rendition of a book titled A Celebration of Poets: Showcase Edition. I don't know if it ever made it to print or became a bestseller, but I at least like to think that I may have a poem setting inside a large coffee table book somewhere in the world. As you read the poem, keep in mind that the Grandfather in the poem represents God.

Spirit Walk

While trekking over virgin snow, above ten thousand feet
The spirits, flying on the wind, will come to you to speak.

The crow, the bear, the lioness,
the wolf, the eagle cry,
Your Manitou, if you are clean, can hear them in the sky.

Their mournful song, it leads and guides you on your way,
and if you choose to listen, from your path, you'll never stray.

Oh Grandfather, keep me strong,
keep me righteous and true.
All I am, I know, is all because of you.

Another poem I wrote reflects a time when I was feeling some regret after having retired from the Army and was no longer serving the nation in what I felt was a useful way. This was before I began working as a Defense Department contractor, where I supported various military and homeland security missions.

It was also before I watched my own children get married and have children of their own. My life became much more interesting and satisfying as I continue to watch my family grow and grow up.

However, at the time of writing, the following poem shows a tinge of darkness, but that little bit of darkness has now been replaced by light. I wrote this sometime between 1999 and 2000.

Old

I lived the warrior's life, I rode hard and carelessly.
Jumped at the face of death when'er his head I'd see.

The rifle was as much my friend as faceless memories.
Craziness invades the space where once I could run free.

Some young men live their life never planning to grow old.
Young and bold's a winning hand, but seldom turns to gold.

I look in the mirror and see my father looking back at me.
I never planned on growing up, how's that reflection me?

I'm years and years older now, than I ever thought I'd be.
Life can be good or cruel at times, look what it's gave to me.

How does a runner slow to walk, when the world's a speeding clock?
Every day I live, I get more behind, the speed of life can blow your mind.

I believe a man should mean just exactly what he says.
That's the way I lived my life while in the Green Berets.

In a world full of leaders, today there are few heroes.
I'm ancient in my thoughts, people say I just don't know.

The real world isn't black and white; it's all just shades of gray.
Many have no room for honor, or the old American way.

I'm years and years older now than I ever thought I'd be.
Life can be good and cruel at times, look what it's gave to me.

A Little Help from My Family & Friends

This section of the book is where I have included some short commentary or stories that I have not written, but were written for me, either about or to assist with my book. I share them as written, as a special addition to my collection of memories I'm passing along.

A Review by John Easterling

I read each word with interest and awe that I had a son that was so very inspirational to so many. Your experiences hold a 007 mystique and many civilians who will read this will have no comprehension of the importance to our National Security your life has made. Some men play with some shape of ball and make millions. Some have their makeup put on by a beautician before they go in front of a camera. They are brave and do daring things, but it really is a stunt man that does the deed. You, on the other hand, have talked the talk and honorably, walked the walk. You are among the unsung heroes that make and keep America the Nation she is. You have changed uniforms several times, but you have never changed the spirit that lives within your very soul. You make me proud.

Notes from Josh on his Mission with SFODA-336, 1992

This was provided by my son, Josh. His memories of the event, while a bit different than mine, are so awesome because they are the memories of an 8-year-old, being recalled 28 years later. Further demonstrating, what we do in life, echoes in eternity.

By Joshua J Easterling

I remember a few bits and pieces of that trip. I remember the ride down, and you let me get the biggest bottle of some sort of new fruit/sugar juice I could find at a gas station. It wasn't good, but I felt cool drinking it in an Army truck. I remember getting there and was having a briefing when one of the guys grabbed the M60 and chased another guy up the stairs of a mocked-up building ripping off rounds. I thought "now THAT is horseplay".

I remember you teaching me how to shoot the 3-round burst and being glad the "kick" wasn't bad...and that machine guns are just as cool as I thought they would be. We had a night op and I remember you guys set me up by a pile of debris next to a 3 or 4-foot ridge. I think I was with you and another guy. There was a tree line probably 200 meters out that we could barely make out and you guys were talking about how the fire was probably going to come from that direction, so I silently fixed my sites there.

Not long after, a few helicopters flew by and there was another sound that wasn't quite like helicopter rotors. Your buddy said something like "those sonofabitches timed their attack with the birds" and then, it was on.

I was Rambo gripping the pig and letting it rain on anything I saw move. I remember watching you shoot and move and thought "whoa...I haven't seen that stuff from Dad." But, you were smiling, laughing, and talking smack, so I knew it was ok to have fun.

I barely remember cleaning, I think I ended up doing some coloring or something, and a ride home being tired but feeling pretty damn tough. I wish I could remember more, as I'm sure a lot of cool things happened, but I remember how it made me feel. Hope this helps!

Tomato Dirt

This story was written by my dad about an outing we had while we were living at Ft Knox, KY around 1966 or so.

By John W Easterling

While I was trying to remember some of the short time that we got to spend together, my mind wandered back across the years to a time when we were out at one of the most questionable places that a Father and his young son could ever want to see. The curing beds below a Sanitary Sewer Treatment Plant holds many things, but as a tourist attraction, it lacks in most venues.

On this day, your Father was in search of young Tomato Plants. There are some things that the amazing human digestion system does not digest and doesn't have any effect on the miracle of the Mustard seed or on the Tomato seed. Some of the largest and most tasty Tomato plants that your Dad had raised in the front yard garden of his Military Housing Area home was from seeds that had passed through someone's bowels and managed to wash up alongside a curing bed where the moisture left the excrement of a million up-stream dumpers.

Combined with the moisture in such a revolting mass of flotsam and touching the well-fertilized edges of a manure curing bed, the seeds found a home and grew into tall, stout Tomato Plants that any Roadside stand would be proud to sell to aspiring young gardeners. This day was like all others and the plants were thriving. You in your exuberant youth was enjoying the outing and although your Father had explained to you that these were curing beds and that they were full of human waste, that you should never, never throw rocks into the beds as they would someday be dry and the remains would be used for fertilizer and they didn't want it to be full of rocks.

Now to an adult mind, that made a lot of sense but to an 8-year-old boy kid, it was wasted breath and you continued hopping and skipping along with your Dad on the quest for the perfect

Tomato Plant. As you went along enjoying the degree of freedom of being several feet in front of your Father, you would clandestinely pick up a rock now and then and toss it into the muck where it didn't really make a splash, but more of a glurp.

You came across a rather large rock. By large I mean about the size of your head and it took both hands to pick it up. With careless abandon, you staggered over to the very edge of a muck pit and heaved the rock into the mire. The result was again a resounding "glurp" and followed by a rapidly rising glob of curing excrement that by chance or perhaps by design, headed directly toward your amazed face. What thoughts went through your mind in that micro-second may never be known, but it had to be something like "OH NO!"

Of course, we had no clean water and a total absence of towels or rags, so your Father took the lesser of two shirts (your "T" Shirt) to clean some of the effluent from your innocent face and having already selected several healthy plants, we headed for home where your mother had you take a shower to be sure you were sterilized from what was to be known in the future as your Shitty, Shitty, Bang, Bang experience.

Intro to Coon Story

This was written for my use by my dad after I had already written about the same incident earlier. As it shows the same story from a different perspective, I wanted to include this version as well. It being as accurate, if not more-so, from his recollections via second-hand information as mine would be, as I was only five and probably only remember what I've been told anyway.

I Ain't Sleeping Out There with No Coons!

By John W Easterling

A short blivet for your Biography should you chose to use it. I must tell it in the third person since I was in Korea when this incident in your young life happened. It was while you and your Mother and Cathy were living on Richland Avenue. As I recall, it was a nice house with a large backyard and edging out behind it a large wooded area. Admittedly it was probably full of Lions, Tigers and Bears, Oh My. At least to a young explorer that was setting out to conquer lands yet unknown to modern man.

This intrepid traveler spent most of the first day setting up his base camp where with many trips from the safety of the Home Front, he transported the supplies and victuals to sustain him during the long journey that would keep him far from the safety of four sturdy walls with doors that when shut would deter all denizens that only came out at night. He was not without experience in these types of adventures. He had on numerous occasions spent nights in a tent along with his Father in the past.

His Base Camp consisted of a sturdy tent with a zippered door flap that would keep him secure at night. He had carried several of the comforts from the Home Front that he would need to sustain life while out there on the edge of a wilderness yet untouched by human footprints. It was enough to instill fear in the bravest of hearts, but this fearless conqueror of the wildest of wildernesses knew no fear.

What? Me? Afraid? I think not. The sun was starting to draw down and the wilderness in the distance was taking on a dark and sinister look that was not there while the beams from a setting sun was still placing sunshine through the trees. It was at about that time when a friend of the Family he called Grampa mentioned something about the raccoons that had been raiding the garbage cans on the back porch at night.

He was just carrying on meaningless conversation when he also alluded out loud and in the presence of our heroic explorer of distant lands that those raccoons probably lived right out there in that forest where the base Camp had been established taking our hero most of the entire day. Blankets, pillows, stuffed animals to keep him company during the long night. Of course, there was the several bags of assorted chips, the sandwiches his Mother had lovingly packed for him and three cans of Coke to quench his thirst while away from civilization. Oh yes, and a flashlight just in case he might need to check out the sounds of night.

These included the scratching sounds of Crickets, the wailing of the Cicadas, the lonesome sound of the Whippoorwills calling to each other in their endless song and the crunch of leaves as the wolves paced around his tent. There was also an occasional screech from an Owl that sounded not unlike the wail of an insane person in pain. Yes, a flashlight was an absolute necessity.

Just as it was going well into twilight and right before true darkness set in on the trail, His Mother looked out into the yard and saw trudging toward the house, the explorer, loaded down with the supplies he had spent most of the day lugging out to his Base Camp Tent. He had it all wrapped up in his arms with the blankets dragging along behind him and a grim look of determination on his haggard face.

When his Mother kindly asked him what he was doing, his answer has since become a phrase used by most of us that remembers the incident, which by the way was told over and over again every time two or more of the Family or friends have gathered since. His words needed no translation or interpretation, they were simply: “I ain’t sleepin’ out there with no coons!”

From My Grandchildren

Kody's Story

By Kody McDowall, 9 years old

Me and my Grandpa have been very close since the beginning of my life in 2010. He and I have grown even closer for about nine and a half years. We grow closer almost every day! We grew much closer as he taught me to play chess and other games. He and I also enjoy knives and rifles together. We have grown closer with that. Also, he and I enjoy ball and Frisbee and other outdoor games. I always enjoy golfing with him, Kitten, and Grandma. Although I have other friends, Grandpa has always been the closest. In this writing, I have tried to illustrate the friendship of me and Grandpa.

Kerrington's Story

By Kerrington McDowall, 7 years old

Me and Grandpa love to swing together. Me and Grandpa like to watch TV together. Me and Grandpa like to play together and build together. We like to read at night together. The last book we read was *The Five Children and It.* That was my favorite book he has read to me. I love Grandpa.

Avalyn's Story

Avalyn Easterling, 11 years old

A memory that I like is when Grandpa was telling us a story at bedtime about lightening. He told us it had something to do with Thor.

Harper's Story

Harper Easterling, 9 years old

I like it when Grandma makes Mac and Cheese for dinner and Grandpa says it's frog legs or something gross like that. Or when he pulled a pole out (of the ground) and he flew backwards!

Sam's Story

Sam Easterling, 6 years old

One thing I love about Grandpa is that every time we have dinner and there's a cover over our dinner, Grandpa always says it might be goat bones or frog legs.

Lori's Memories with Dad

By Lori McDowall

I have so many memories about my Dad, it has taken me a few weeks to think through the ones I want to include here. I will highlight the few that really standout, for good or bad reasons (ha!).

One of my earliest memories of my Dad was the day he and I made a snowman outside of our trailer home in Albany, OH. In my mind, it was a very large snowman with a hat. I remember my Mom telling me the next day that when she came home that night from work, she thought the snowman was a man standing in our yard, and she was too scared to drive all the way up to the home until "the man" didn't move for several minutes.

I also have a memory of getting in big trouble when I walked to the pond on our property with my cousin, Chris. Chris was a few years older than me, and I thought it was safe to go with him around the property. Apparently, we were gone quite a while because when we were walking back to my home, my Dad and Mom were hunting for me. When my Dad saw me, he took a ~~really small~~ LARGE "switch" off the tree closest to him and spanked me with it. To this day, I don't understand why I was spanked and Chris wasn't, but I have never felt comfortable around ponds since…

I also remember my Dad picking me up from the babysitter's apartment in Albany. I remember the motorcycle he'd pick me up on, but I don't remember actually riding on it. I also remember going to the "carry out" with my Dad to get beer. I would also ride with him on his horse when we'd go out riding with friends on the back roads.

My next memories are in Germany. My absolute favorite memory of all time with my Dad was when he'd take me for bike rides. I would sit behind him in my "bike seat," and he allowed me to do all the turn signals for our entire ride. I always felt so important and special doing those. I can still remember the smell of that "bike seat" and vaguely remember how excited I would feel when he said we were going to go for a ride. There was a little house across the street from our apartment that had the cutest little trolls in their yard. I loved riding past and looking at them.

I remember eating a lot of steak in Germany sitting around our small kitchen table. That steak is why I can't donate blood now. Apparently, I may have mad cow disease or something. I also remember riding on the trolley to go shopping with my parents, going out to the club with them and getting the best soup I've ever had, and walking around downtown during Christmas time and seeing the angels up in the windows of the buildings. I think I was in my Brownies uniform at the time.

Back in America, my next few memories are scattered. I remember my Dad giving me my first bike at Aunt Debbie's house. It was pink and looked so big! He taught me how to ride it that day. I also remember him "running" me to school in Fayetteville one morning before we moved into quarters. Apparently, I was late, and in my memory, he grabbed my hand and started running. He ran, and I soared behind him like a kite all the way to school. After we moved into quarters, I remember him visiting me at the end of the school year in Mrs. Cole's 2nd grade class. I was SO PROUD to be his daughter when he walked in the classroom.

There are scattered memories from then on, like going to Myrtle Beach, going to Ohio to visit family, etc. However, the memory that stands out the most during this time was when my Dad left for Desert Storm. I remember him waking me up in the middle of the night after he was all packed and ready to leave. He hugged me and said he loved me and would be back soon. I remember crying the rest of the night. I stood very tall and proud when my junior high school played "God Bless the USA" over the loud speaker during a pep rally a few weeks later.

Fast-forwarding several years, I remember watching The Last of the Mohicans with him and thinking about the movie every time we visited the Smoky Mountains. He was always my biggest (and loudest) cheerleader at karate and Taekwondo tournaments. I felt invincible when he cheered for me, and I always fought my heart out to make him proud. The $0.10 hamburgers he used to purchase for Josh and I after practice were always great too.

After we left Ft. Bragg and ended up at Yuma Proving Ground, it was a blur of a year. It was my senior year of high school and full of fun memories, but one memory that stands out was when my Dad and I rode ATVs through the desert just outside of YPG.

While in Colorado, he and I used to get up early in the morning and go power walking before work. On one particular morning, we were just starting to walk out of the garage around 5:00 am, and a hearse began to pull away from our neighbor's house. Without any hesitation, my Dad immediately came to attention and slowly saluted the hearse as it drove away. Our neighbor had served in the military, and my Dad gave him the final honor and respect he deserved. I can only imagine what the driver of the hearse felt at that moment. I was so proud of my Dad for his example to me that I couldn't speak for several minutes.

There are many other great memories after this, like my Dad being at the Peterson Aero Club when I took my first solo flight (I was determined not to crash and die since he was there watching), being on the flight line with me when I was sworn into the United States Air Force, being in Knoxville, TN to see me get my commission, and giving me away at my wedding. However, nothing has compared to the memories I have gathered over the past 10 years watching my Dad be a Grandpa to my children. He loves them completely. He accepts them fully. He always has all the time in the world for them. My children will always know how it feels to be deeply loved by my Dad and Mom, and for this, I will forever be grateful.

Additional Background Information

The next few pages present a listing of my major assignments and operations I conducted as a Green Beret. Most of this has been discussed in the series of stories already presented, but this is more a chronological list, to include some more specific details about my time in Special Forces.

October 1986 – May 1987

Special Forces candidate assigned to the 1st SWTG(A) at Ft Bragg, NC

During this time, I attended Phase 1, 2, and 3 of the Special Forces Q-course. I broke my right leg during the infiltration jump in phase 3 and had to return 6-weeks later and start the 3rd Phase over. I made another jump into the same DZ for the 2nd attempt, which was successful.

June – December 1987

Assigned as a SF Engineer (18C) with SFODA-591, 5th SFG(A) at Ft Bragg, NC

During this period, I performed duties as the self-help materials NCO attached to the Group Engineer section. These duties included designing the plans and the Bill of Material to build the first dust-free, oxygen-safe work area, required for maintenance of the Draeger Lars (re-breather) diving equipment for the Scuba Locker. I also deployed on a Battalion readiness evaluation with a scout swim Special Forces Operational Detachment – Alpha, or SFODA. I participated in the exercise named Last Claw, conducted at Ft Stewart, GA, where I was a member of a Special Forces Scout-Swim detachment that conducted a direct action demolitions raid mission.

January 1988 – October 1989

SF Engineer (18C), at SFODA-594, 5th SFG(A), Ft Bragg, NC

This period I performed duties of an SFODA Engineer NCO Supervisor. I conducted operations in unconventional warfare, foreign internal defense, direct action and special reconnaissance. The Detachment was a Military Free Fall team. I participated in the following exercises:

FTX Cab Acre I 1988, conducted in El Centro, CA, where we trained in Special Forces Military Free Fall (MFF) High Altitude High Opening (HAHO) infiltration techniques.

FTX Ebony Wood 1988, conducted in the vicinity of Amman, Jordan. We conducted infiltration techniques and light infantry training and instruction to elements of the Jordanian Special Forces.

FLINTLOCK 1988, conducted in Kamina, Zaire. (Zaire became the Democratic Republic of the Congo or the DRC, in 1997). We conducted MFF and light infantry training and instruction to elements of the Zairian Army.

FTX Bondeko 1988, conducted in Kinshasa, Zaire, where we conducted MFF and light infantry training and instruction to elements of the Zairian Army.

FTX Cab Acre II 1989, conducted in Yuma Proving Grounds, AZ. We trained in Special Forces MFF HAHO infiltration operations. This type of training was also known as Stand Off Parachuting.

FTX Celebrity Charge 1989, conducted at Ft Huachuca, AZ. The exercise canceled midterm and we redeployed to Ft. Bragg, NC, due to a helicopter crash which killed several members of SFODA-591 in my company during the exercise. The 18C that was killed in the crash had taken my old position after I had been assigned to SFODA-594, the Company's MFF SFODA. I was one of the Honor Guards who attended his funeral in the aftermath.

SOCOM Validation 1989, at Ft Bragg, NC. Our Battalion conducted MOS (job) and cross-training operations, infiltration techniques and direct-action operations. This event was a Congressionally-mandated evaluation of the Special Forces line units in order to justify the unique support requirements (budget and otherwise) and consisted of multiple high-stress and physically-demanding events back-to-back over the course of a week.

MABOTE 1989, conducted in Kamina, Zaire. We conducted MFF and light infantry training and instruction to elements of the Zairian Army.

ARTEP Evaluator 1989, performed at Camp Lejeune, NC. I was tasked to evaluate the performance of an SFODA from isolation to End of Mission and provided the After Action Report, or AAR to the Battalion Commander.

I-TOW Mobile Training Team Mission 1989, at Gaborone, Botswana. I was the MTT Team Sergeant and primary instructor of a four-man team, training and instructing elements of the Botswana Defense Force (BDF) on the Improved TOW (anti-tank) system. Training and instruction ranged from individual operation to strategic tactics of the unit. Also, I was required to interface with personnel from the US Embassy and VIP's, to include the Commander-in-Chief of the BDF. This was an initial entry mission for US Forces into the country of Botswana.

October 1989 – November 1990
Assistant Operations and Intelligence NCO, SFODA-594, 5th SFG(A), Ft Bragg, NC
During this period, I was assigned as the Detachment's Intel NCO. I also performed the duties of the SFODA Warrant Officer or Team Tech during isolation after May 1990. I participated in the following operations:

FTX Cameroon Latch 1990 at Ft Bliss, TX. I was the primary instructor for the 3rd Battalion's Weapons Sergeant, (18B) training on the TOW weapon system, which culminated with a successful live fire exercise. I also conducted desert survival and navigation operations, along with a direct-action live fire. My SFODA also conducted MFF operations.

FTX Cab Acre III 1990 at Yuma Proving Grounds, AZ. I participated in SFODA MFF HAHO or Stand-Off infiltration operations.

SOCOM Certification 1990 at Ft Bragg, NC. My unit conducted MOS and cross-training operations, infiltration techniques and direct-action training. This was a continuation of the SOCOM Validation exercise conducted in 1989.

FTX Ocean Venture 1990 in Antigua, West Indies. My SFODA conducted a light infantry FID mission, teaching advanced level infantry tasks during phase I of the exercise. Phase II consisted of a force on force with the Antigua-Barbados Defense Force (ABDF) and the United States Army 75th Ranger Regiment.

This entire mission afforded training in FID, UW, link-up ops, Counter Narcotic and MFF Infiltration techniques.

FTX Caisson Cowboy 1990 at Helena, MT. I performed the duties of the SFODA Warrant Officer or XO during this operation. Training included MFF and air mobile infiltration techniques. Direct Action and Special Recon missions were performed during the operation.

December 1990 – July 1991
Team Sergeant /Operations NCO (18Z), SFODA-334, 3rd SFG(A), Ft Bragg, NC. This period covered the following operations:

Operation Desert Storm, Kuwait and Saudi Arabia
I began this period as the SFODA Operations Sergeant. This entire period consisted of planning for, deploying to, conducting and finally redeployment from Operation Desert Storm, or the Gulf War. As Operations Sergeant, I also conducted the duties of the Detachment' SF Warrant Officer since the Team's XO was reassigned to the BN HQ. These duties included the supervision and direction of team training geared towards Special Recon and Direct-Action. The combat missions were Direct Action operations, followed by security assistance after hostilities officially ceased. During Security Assistance operations I worked as the G2 officer counterpart of the Kuwaiti Northern Area Command (Division Level).

August 1991 – May 1992
1st Battalion S2 NCOIC or BN Intelligence and Security (18Z), Ft Bragg, NC
I spent this period as the S2 NCOIC. Working in this position allowed for experience in a staff role, as well as gaining insight into the Intelligence community. During this period I supported the following operations:

FTX GRUFF RIDER 1991, at Camp Lejeune, NC, where I deployed with Forward Operations Base (FOB) and worked as night shift OIC providing Intel and daily briefs to the BN

Commander, and supervising the attached USAF weather team.

GRIDIRON SCOOTER II, January 1992, at Ft Bragg, NC where I participated in 3rd SFG(A)'s 1st USASFC VALEX as the FOB S2-OIC.

May – October 1992
Assistant Operations / Intelligence NCO (18F), SFODA-336, 3rd SFG(A), Ft Bragg, NC
I worked this assignment as the Assistant Operations and Intel NCO on a Scout Swim SFODA. Missions I participated on include the following:

Counter Narcotic JTF-6 at Ft Bliss, TX, where my Detachment performed water ops and various other training, while preparing JTF mission Intel briefs and a site survey for a proposed JTF mission site. Participated in multiple Counter Narcotic missions.

Joint Readiness Training Center (JRTC) at Ft Chaffe, AR where my SFODA performed FID and Unconventional Warfare missions with a guerilla force made up from Army IRR personnel. Trained company sized unit in light infantry tactics and techniques, culminating in a live fire ambush. Prepared Intelligence Analysis and ensured all intelligence requirements were met during the mission.

01 November 1992 – 14 February 1994
Assistant Operations / Intelligence NCO (18F), SFODA-384, 3rd SFG(A), Ft Bragg, NC
During this period, I held the position of the Assistant Operations / Intelligence NCO for an MFF SFODA. Upon assignment, I began to focus on the upcoming Battle Focus Assessment (BFA) requirement and the upcoming deployment to one of the target countries to conduct static line and free fall cross training with host nation counterparts. I also assumed the duties of SFODA Warrant Officer as required during all missions which included the following:

Lancer Jaguar 1993 at Harare, Zimbabwe, where we conducted a 3-team initial entry mission into Zimbabwe, a former Soviet allied country. The operation was a very high visibility mission and training was visited by more than 10 different countries Defense Attaché Officers (DAO) and key civilian political leaders. Training was given in static line parachuting, demolitions, drop zone/landing zone operations and light infantry tactics. Training received by the SFODA from the host nation included combat-tracker techniques (formerly Rhodesian Tracking School), host nation airborne ops, and Zim infantry tactics. Additional training was conducted in MFF, both HALO and HAHO, static line parachutist Jumpmaster duties in non-standard winged and rotary aircraft, and care, maintenance and packing of various personnel and equipment parachutes. Aerial delivery training with A-series bundles was also accomplished.

Present a Static Line Jumpmaster Course at FT Bragg, NC during June 1993 to train members of the Special Operations community in Jumpmaster and Drop Zone Support Team operations. I was in charge of the pre-mission operations and coordination/taskings. During the conduct of the course I remained the Operations NCO and then drafted the After Action Report upon completion. During jump week I was primary Jumpmaster for the five Actions in the Aircraft training jumps and five MFF jumps. The mission was a success, culminating in the graduation of 58 qualified Jumpmasters out of 67 initial students.

Joint Readiness Training Center, FT Polk, LA, August and September 1993 to conduct the planning and infiltration into a denied area for a Special Reconnaissance mission. The target was a suspected chemical training and storage area. This mission required an additional three chemical specialists to be attached to and infiltrate with the SFODA. Also, it required extensive planning and training for employment into chemically contaminated areas requiring portions of the mission to be conducted in various levels of protective equipment.

Counter Narcotics, JTF-6 in the vicinity of Big Bear Lake and March AFB, CA to conduct counter-drug missions in the national and state forests near Los Angeles, CA. This real-world mission included "gray" infils and exfils as well as clandestine commo plans to prevent compromise of the activities going on in the Area of Operations (AO). During this mission there was no officer with the team and I acted as the SFODA Warrant Officer (XO), as well as the Intel NCO.

February – June 1994
Team Sergeant / Operations NCO (18Z) at SFODA-384, 3rd SFG(A), Ft Bragg, NC
I spent this period as the Operations NCO for the MFF SFODA I was already assigned to, having been assigned to the Team Sergeant position after the former Team Sergeant retired. Missions included the following:

Falcon Cherokee at Harare, Zimbabwe from February to April to conduct a mission that was a continuation of the operation conducted there during the previous year. The primary focus was the continuing introduction of the steerable static line parachute system to the host nation's Para's, SAS, and Commandos. Additionally, other areas covered were aerial delivery, (A7-A, A-21, A-22 bundles), Pathfinder operations, demolitions training, and weapons familiarization. Also, MFF operations were conducted with the Zimbabwe Air Force and SAS. This operation also had a dedicated USAF MC-130H aircraft for one week of training. During this week, the first night time static line drop of Zimbabwe Army paratroops from an American aircraft was accomplished. The message pick-up system (MPU) was also introduced to the host nation Army and Air Force.

Agile Provider 1994 at Camp Pendleton, VA. This mission was to provide support to the Battalion HQ. I conducted duties of an Area Specialist Officer (ASO) and was responsible for supervising and coordinating the support of multiple SFODAs in isolation conducting mission planning.

June 1994 – June 1995
Senior Military Free Fall Instructor at the Military Free Fall School (MFFS), FT Bragg, NC
I began this period as a MFF instructor candidate. I was awarded my MFF Instructor Rating (MFFI-336) at El Centro, CA. in February 1995. The MFFS ran three basic courses at

NAF El Centro as practice for the upcoming move to Yuma Proving Ground, AZ.

June 1995 – May 1996

MFFS Company Operations NCO / MFFS Instructor at Yuma Proving Ground, AZ

I completed this period as the Operations NCO for the Company during our Pilot course and subsequent courses at YPG. This was a special time due to all the different areas I was responsible for, to include, building construction, contracts and purchasing, establishing working connections with the school and other elements on base, plus the normal operations work of not only YPG, but also the Company and Battalion personnel back at Ft Bragg. I also performed duties as a MFFS Instructor with the Basic and Advanced MFF courses. I was also the Company Contract Officer Representative (COR) with the responsibility to order and inspect services provided by the contract air company flying in support of the Military Free Fall School. During the period of 24 Feb to 07 Apr 96 I served as the Company First Sergeant. During this time, I was responsible for the normal daily activities of the Company First Sergeant, as well as training the new Operations NCO in the duties of his position. During my time at the MFFS, I also completed the Military Tandem Vector System training and was designated as Military Tandem Master #17. Additionally, I was promoted to Master Sergeant (E8) during this period in August 1995.

June 1996 – September 1999

Team Sergeant / Operations NCO, SFODA-061, 10th SFG(A), FT Carson, CO

I completed this assignment as the Operations Sergeant of SFODA-061, which had an additional duty as the 10th Group's Special Operations Target Interdiction Committee (SOTIC). This was a unique assignment due to the requirement to not only maintain currency in all areas as an SFODA, but to conduct the 10th Group's sniper sustainments, basic sniper courses, host the annual 10th Group Sniper Competition, direct and overwatch the sniper weapon systems maintenance plan, and provide the force with research and development of current and new sniper weapons and related technology. Operations and missions I participated in include the following:

Sniper Sustainment 1-97 at Ft Carson, CO, where my Team planned and executed annual sustainment training for ten SOTIC qualified soldiers from 10th SFG.

Robin Sage (SF Q-Course) support at Camp Rowe (Mackall), Ft Bragg, NC where I performed the role of the Commander of a "guerrilla" force for the Robin Sage Unconventional Warfare exercise run at the culmination of training for Special Forces students attending the SFQC.

4th Annual International Sniper Competition at Ft Carson, CO for which my Detachment planned and executed the annual competition. Competitors were from all branches of Special Operations units, conventional US Army, US National Guard, and various foreign units.

Extreme Cold Weather Training (ECWT) at Camp Hale, CO during January 1997. This 10-day period in ECWT was completed in temps not exceeding zero degrees Fahrenheit and at elevations of 9400 to 11500 feet above sea level. Training included cross country movement on skis and snowshoes, as well as skijoring. Snow cave construction and general Arctic environment survival skills were also practiced. Tactical movement and combat drills were also conducted. Long-distance movement operations, with and w/o equipment sleds, was performed.

A Counter-Sniper Commander's Seminar at Grafenwöhr, Germany was conducted by myself and my Intel NCO during this quick-turn trip. We conducted this three-day trip for a one-day briefing as a pre-Bosnia-Herzegovina deployment training requirement for the 1st Infantry Division's command elements. The audience was Company Commander and above.

Sniper Sustainment 2-97 at Ft Carson, CO. My Detachment planned and executed sniper sustainment for eight SOTIC qualified soldiers. The Team also provided sniper weapon familiarization classes to six non-SOTIC trained individuals.

This was to train them in care and maintenance of the Sniper Weapon System for upcoming missions.

Pre-Deployment Training for Bosnia rotation at Camp Guernsey, WY
My SFODA deployed to Camp Guernsey to practice Detachment live fire combat drills for both mounted and dismounted scenarios. This was in preparation for the upcoming Joint Commission Observer (JCO) mission in Bosnia-Herzegovina. We also trained with live fire and dry fire with Close Air Support via MC-130 aircraft and other aerial platforms. During this deployment I was the acting Company Sergeant Major.

Operation Joint Guard in Bosnia-Herzegovina from March through August 1997
My SFODA Commander and I arrived in Tuzla, Bosnia-Herzegovina (BiH) on 31 Mar 1997. After initial in-country briefs, we traveled to Vares (Federation). We performed preparational duties to support the rest of the Detachment's deployment. After two months in Vares as JCO F10, we relocated to Vlasenica in the Republika Srpska area of BiH. We deployed here as JCO N70. This mission demanded exceptional people and various ASOT skills. Our team of six men lived in the town with the local population.

SOTIC CAT II Course at Ft Carson, CO, where the Detachment planned, prepared and executed a Special Operations Target Interdiction (Sniper) course for soldiers from 2nd BN 10th SFG(A). This was the first 10th Group SOTI course that had been conducted over in 24 months. The course produced individuals trained in advanced marksmanship and fieldcraft skills, able to engage selected targets from concealed positions at ranges and under conditions that are not possible for the normal rifleman. Actions required were supervision and coordination for vehicle and personnel support, ammunition forecasting, range support taskings, training concept, operations order and after-action report writing. I also served as the Detachment Commander during the course.

5th Annual International Sniper Competition at Ft Carson, CO during January 1998. The SFODA planned and executed the annual competition. Competitors were from all branches of US Special Operations units, conventional US Army (Active Duty and Guard), and foreign NATO units.

Battalion Train-up for Bosnia at Camp Guernsey, WY and Ft Carson, CO during March 1998 where I conducted planning and coordination as the NCOIC of the Command and Control cell of the Battalion's Pre-Mission Training for Bosnia & Herzegovina. My SFODA deployed to Camp Guernsey as part of the advance element and finalized arraignments for billeting, ranges and other support requirements. I continued as Command and Control during Ft Carson phase, after bad weather forced training to be canceled at Camp Guernsey. Areas of training included Close Air Support, Close Quarters Combat, Defensive Driving, NBC (contaminated environment), Military Operations on Urbanized Terrain, and Advanced Marksmanship training.

Military Retirement at Ft Carson, CO on September 30, 1999 although I began terminal leave around the end of June 1999. I had planned to stay in the Army longer, but decided to retire after being told I had to leave my position on the SFODA and start working as a Staff NCO filling a Company or Battalion HQ level position. I believe retiring at that point turned out to be the best choice I could have made.

At the time of my retirement from the US Amy, I held the following certifications and awards.

Military Occupational Skills (MOS)

18Z5HW8 Special Forces Senior Sergeant - Instructor (H), MFF (W8)

18F4HW8 Special Forces Intelligence Sergeant, Secondary MOS

18C4OW8 Special Forces Senior Engineer Sergeant, Additional

11B3P Infantryman, Parachutist, Additional

13C3P TACFIRE Section Chief, Parachutist, Additional

13E3P Fire Direction Control, Parachutist, Additional

Awards and Honors

Combat Infantryman's Badge

Meritorious Service Medal (MSM) - 3 awards

Army Commendation Medal (ARCOM) - 5 awards

Army Achievement Medal (AAM) - 4 awards

Good Conduct Medal - 5 awards

Kuwaiti Liberation Medal (SA & KU)

NCO Education System Ribbon w/#3 (PLDC, BNCOC, ANCOC)

Army Service Ribbon

Overseas Ribbon #2 (2 assignments to Germany)

Special Forces Tab

Master Parachutist Badge

Master Military Free Fall Badge

1st Special Forces Regimental Crest

Special Forces Combat Patch

Foreign Awards

Zaire Military Free Fall Wings

Jordanian Senior Parachutist Wings

Zimbabwean Army Parachute Badge

Zimbabwean Air Force Parachute Jumping Instructor Badge

Zimbabwean Special Air Service (SAS) Wings

Military Schools and Courses

Special Operations Target Interdiction Course (Level 1), 6-weeks, 1996
Advanced Military Free Fall/Tandem Master, 2-weeks, 1996
Advanced Military Free Fall Parachutist Course, 12-weeks, 1994
Instructor Training Course, 2-weeks, 1994
Advanced Special Operations Techniques Course, 12-weeks, 1992
Special Forces Engineer Advanced Noncommissioned Officers Course, 12-weeks, 1991
Special Forces Operations and Intelligence Course, (MOS 18F) 13-weeks, 1990
Ranger Training Brigade Light Leaders Course, 4-weeks, 1989
Advanced Target Analysis Course (CIA), 2-weeks, 1989
Military Free Fall Jump Master Course, 3-weeks, 1989
TOW Trainer Course, 4-weeks, 1988
Advanced Military Free Fall Stand-off (HAHO) Course, 3-weeks, 1988
Military Free Fall Course (HALO), 5-weeks, 1988
Special Forces Engineer Sergeant Qualification Course, (MOS 18C) 21-weeks, 1987
Static Line Jumpmaster Course, 2-weeks, 1986
Basic Airborne Course, 3-weeks, 1985
TACFIRE Fire Support Course, (MOS 13F) 14-weeks, 1984
Artillery School Operations and Intelligence Course, (MOS 13E) 12-weeks, 1975
Basic Combat Training, 8-weeks, 1975

Date of Latest Rank

01 SEP 95 (Master Sergeant, E8)

Security Clearance

Top Secret

Additional Information
Silas Easterling and the Skirmish at Hurricane Bridge

The hunt for information at the "Skirmish of Hurricane Bridge" as it became known, led me to the following poem. The 13th VA. VOL is actually a Union (Yankee) unit from West Virginia. Many of the "West Virginians" still thought of themselves simply as Virginians since WV didn't become a state of its own right until 1863.

So for some needed explanations, Albert G. Jenkins was the Confederate (Rebel) General who was leading the Confederate States of America (CSA) elements that attacked the Union forces at Hurricane Bridge. He sent an officer to the Union camp and offered to let them surrender, they chose to fight, and the battle was on. Johnson was the Union Captain in charge of the Yankees holding the bridge, and who told the CSA officer that they would not surrender, but fight instead.

The War of the Rebellion: A Compilation of the Official Records of the Union and Confederate Armies. Washington: GPO, 1897. Series I, Vol. 51, pt. 1, p. 176.

MARCH 28, 1863. - Skirmish at Hurricane Bridge, W. Va.
Report of Capt. James W. Johnson, Thirteenth West Virginia Infantry (US).
HURRICANE BRIDGE, VA., April 3, 1863.
SIR: I have the honor to make the following report of the assault made on this post by the rebel General Jenkins and his command on the 28th of March:
About 6 o'clock in the morning of the 28th, our pickets brought in a flag of truce, with the following note from Jenkins (CSA):
Colonel Brown (US),
Commanding Thirteenth Regiment U. S. Volunteers,
Hurricane Bridge:
I have now an overwhelming force so disposed as to completely surround you and cut off your retreat. A humane desire to avert the loss of life induces me to demand your surrender. In the event of your compliance, and the surrender in good faith of all forces under your command, they shall receive the treatment warranted by the usages of war, and both officers and men will be paroled. Twenty minutes will be allowed, for the consideration of this note and to return a reply.
I am, Colonel, very respectfully, your obedient servant,
A. G. JENKINS, Brigadier-General, C. S. Army.

CPT Johnson (US): Upon the receipt of the above note I immediately sent in reply that I should not surrender the forces under my command unless forced to do so by an exhibition of his boasted strength, and immediately set about making the best possible disposition of the limited forces under my command. In fifteen minutes, we were ready for action. All our available forces, numbering about 150 effective men, were drawn up inside our fortifications, when the enemy appeared in force and opened a furious fire upon us simultaneously on three sides from as many different hills, owing to the high elevation of which and the unfinished condition of our works, exposed our men to a most galling cross fire, which they withstood and returned with the firmness of veterans. The enemy's sharpshooters, posted on the adjacent heights and armed with globe-sighted rifles, were constantly endeavoring to pick off officers and men. After about five hours' brisk and animated firing from both sides, the enemy sullenly withdrew his forces, leaving a few of his wounded, who fell into our hands, from whom we have learned that the enemy's force engaged did not number less than 500 men. Our loss was 3 killed and 4 wounded, one of whom has since died. To both officers and men, I return my most sincere thanks for the bravery and gallantry displayed during the engagement. Where so many heroic deeds were performed it would be unjust to mention individual acts of gallantry. It is enough to say that all behaved in the most noble and gallant manner.

I have the honor to be, sir, very respectfully,

J. W. JOHNSON, Captain, Comdg. Detachment Thirteenth Regt. Virginia Vol. Inf. (US)

Col. W. R. BROWN, Commanding Thirteenth Virginia Volunteer Infantry. (US)

Skirmish at Hurricane Bridge

Point Pleasant Weekly Register
April 30, 1863
BATTLE OF HURRICANE BRIDGE!
By M. V. B. EDENS, CO. A. 13TH VA. VOL. (US)
(US – 13th West Virginia Infantry Volunteers)

Between a high hill and circuitous ridge,
Is the neat little town, known as Hurricane Bridge,
Or at least, 'twas a neat little village before -
The rebellion arose in the bright days of yore.

But rebellion has banished those pleasures, so sweet,
And the wail of despair is now heard in the street -
The many fine buildings in ashes are laid
The wealth of the inmates forever decayed.

The 13th Virginia is camped near this town,
Who in a late battle have won great renown.
The regiment is formed of a brave set of boys,
Who accomplish their work without making much noise.

But Albert G. Jenkins not aware of that fact,
Came early one morning to make an attack,
With a yard of white cotton unfurled to the breeze,
His staff came into camp, while Jenkins stayed behind trees.

'Twas Samuels who brought in the banner of truce.
Says Johnson, to surrender I this morning refuse.
In the time which is given to my works I'll repair.
And give you a battle on principles fair.

When cool as an iceberg we fell into line.
To defend the bright stars which on our banner doth shine,
And to hold our position at Hurricane Bridge,
In spite of the rebs on each neighboring ridge.

For Jenkins marched forth with his menacing host
To capture or drive us away from our post,
He threw his men round on the top of each hill
To make us surrender or fly at his will.

He thought with his men on the left and the right,
He would capture our squad without having to fight,
And with Yankee clothing he his men would equip,
And thus would secure successful his trip.

But Albert was much disappointed to find
To surrender that morning we were not inclined
Then he told all his men to prepare for the blow
And soon they'd conquer the insolent foe.

From each hill top they fired on the 13th in vain,
For each at his post did boldly remain.
For the space of four hours he fought very hard
But his incessant firing we not much did regard.

Like Jeff Davis' scrip that's more plenty than good,
They sent in their lead from each neighboring wood
Like the insects that sing in the long days of June,
The balls from the rebs almost whistled a tune.

But is was not sufficient to induce us to dance,
It was like Mason in England or Slidell in France,
Who after they made known their treacherous plan,
Have to end their hard labor just where they began.

With stomach's quite empty and feet without shoes,
Each felt like a bankrupt with a spell of the blues,
Like a dishonored guest ere the bankquet [sic] is o'er
Arrives at the feast and is kept from the door.

They knew in our camp were crackers, coffee and meat,
And they longed to get in to get something to eat.
Like Cruse's man Friday, who on the mountain did stand,
And beheld in the distance, his own native land.

With obtsacles [sic] in front which he could not surmount
Was not able to settle the lengthy account,
And fell ere he crossed o'er the dark rolling wave
And found in the deep sea a watery grave.

So Jenkins after giving us a hard shower of hail
And finding with muskets he could not prevail
Sent out his command to retreat from the field,
For the 13th Virginia was too stubborn to yield.

Being tired and hungry, they left in disgust,
Aware of the fact that Albert won't do to trust;
He promised to give them both food and attire,
In short everything that a reb could desire.

Except a good thrashing they left as they came
With nothing to brighten Jeff Davis' fame;
And Jenkins found out that one blunder he made
In seeking for sunshine he found nought but shade.

Some Additional Hurricane Creek History

"It was a cool, misty morning in the year of 1774. A party of surveyors commissioned by George Washington was traveling down the Kanawha River. As the sun was slowly rising, cutting through the fog of this early spring day, they noted in their journal that all the trees at the mouth of the creek were bent in the same direction... and they called the location "the place of the hurricane".

The creek became known as Hurricane Creek because of this phenomenon, and by 1811, according to early Virginia maps, the town of Hurricane Bridge was located where Route 34 now crosses the creek near Route 60.

About the Author

Seems like yesterday we were in El Centro and you made your cherry blast with the team, a HAHO. I can still see it clearly, all of it, the sky, the plane, the ground, and seeing you. You cut away and I remember seeing you under canopy a long way off and below us all.

- Mike Gaglione, Special Forces (retired)

Bill Easterling is one of the finest NCOs that I served with and would trust him with my life! I'm proud to have served with him. The one regret I will always have is not being there to support him when 3rd Group deployed to Desert Storm.

- Thomas Tannahill, Special Forces (retired)

It's been many moons since our days on SFODA-384 together. One thing I recall is that you were a significant mentor to me as a young Green Beret. One of my early "life shapers" if you will.

- SGM Eric Rasmussen, Special Forces

We had a great time on the Team for certain! Our Sniper School trip, Germany Counter Sniper trip, Bosnia trip, the list goes on...

- Phil Crowley, Special Forces (retired)

Mr. Easterling served in the US Army as a Special Forces "Green Beret" and further, after retiring from the US Army, as a security contractor for the US Departments of Homeland Security and Defense. He spent the majority of his US Army career serving on a Special Forces Operational Detachment - Alpha during periods of peace and combat.

While serving as a Special Forces operator, his primary duty position was as the SFODA Operations and Intelligence expert. While a member of the Special Forces, and more recently as a contractor, he has worked in many geographical locations to include Africa and the Middle East. He has served during combat operations in Kuwait during the 1991 Gulf War, Bosnia-Herzegovina, and Baghdad, Iraq during Operation Iraqi Freedom. Examples of Mr. Easterling's operational experience include the following mission types and locations:

Botswana: Anti-Armor System Training Team

Dem. Rep. of Congo (formerly Zaire): VIP Security and Counter Terrorism

Zimbabwe: Advanced Parachuting, Light Infantry Tactics, Combat Tracker

Bosnia-Herzegovina: Force Protection, International Forces Liaison

Desert Storm, Kuwait: Urban Combat, Explosives Disposal, Civil Affairs

Baghdad, Iraq: Force Command Intelligence Support Management

USA: Critical Infrastructure Protection and Counter Narcotic Surveillance

Made in the USA
Monee, IL
08 November 2026

67388665R00236